I0486361

The Digital Dictionary

*A handy guide
to find explanations
for common technical buzzwords.*

Note for Librarians: a cataloguing record for this book that includes
Dewey Decimal Classification and US Library of Congress numbers is
available from the National Library of Canada. The complete cataloguing
record can be obtained from the National Library's online database at:
www.nlc-bnc.ca/amicus/index-e.html
ISBN 1-4120-3908-8

TRAFFORD

Offices in Canada, USA, Ireland, UK and Spain
This book was published *on-demand* in cooperation with Trafford
Publishing. On-demand publishing is a unique process and service of
making a book available for retail sale to the public taking advantage
of on-demand manufacturing and Internet marketing. On-demand
publishing includes promotions, retail sales, manufacturing, order
fulfilment, accounting and collecting
royalties on behalf of the author.
Book sales in Europe:
Trafford Publishing (UK) Ltd., Enterprise House, Wistaston Road
Business Centre, Wistaston Road, Crewe CW2 7RP UNITED KINGDOM
phone 01270 251 396 (local rate 0845 230 9601)
facsimile 01270 254 983; info.uk@trafford.com
Book sales for North America and international:
Trafford Publishing, 6E–2333 Government St.,
Victoria, BC V8T 4P4 CANADA
phone 250 383 6864 (toll-free 1 888 232 4444)
fax 250 383 6804; email to bookstore@trafford.com

www.trafford.com/robots/04-1716.html

10 9 8 7 6 5 4 3 2

ABOUT THE AUTHOR

Michael Phillips was born in Bonn, Germany, the second son of typesetter Heinz and his wife Hildegard. After completing the normal extensive educational training, he joined the military, serving four years and earning an honorable discharge. Over the next ten years, he held a number of upper management positions at Radio Shack and Alpine Electronics in his home country. He then decided to immigrate to the United States, where he has fulfilled his lifelong dream of becoming involved in the film and movie Industry.

During the course of his first 15 years working with electronic media and technology in the U.S., he frequently would overhear the experts talking, but he, like many apprentices and "newcomers", was often unclear on <u>what</u> those professionals were talking about. Finally, he came to the conclusion that there was something missing, a reference source of some kind, that ordinary people could use and help them understand the tech-talk that increasingly surrounds them.

For weeks he searched bookstores, trying to find a dictionary or handbook that would shed some light on the "mystery terms" the tech experts used. His search was unsuccessful and so, in 1995, he began to write the handbook himself. His goal was to create a pocket-size reference that would allow people to quickly find an answer to the most basic questions and definitions of the terms most widely used in his part of the industry, the vastly expanding communications world and the approaching digital revolution.

He began his project in a very simple way: by writing down any words or phrases that he felt required clarification. For example, most people may know what a DVD is, but do they really know the meaning or just the acronym? And how many of us would like to know what a "jib" or a "gaffer" is? How about a "scratchpad?" A "base station?" A "kilobyte?" And what does a "cache" do? In his spare time, he identified more than 1000 acronyms and terms and researched their origin and meaning...and The Digital Dictionary was born!

ACKNOWLEDGEMENT

If I attempted to acknowledge every person and every source that has helped me in my nine years of research studies to compile this book, it would change the Dictionary's handy pocket size to that of a World Atlas. I have therefore decided to express my gratitude in a simpler, more global way. So, to all those helpful individuals who have crossed my path, I will simply say a heartfelt "Thank You!" I also want to acknowledge the Guiding Light of the Universe for leading me to the right people and showing me where to find the research data to accomplish my goal. My life companion Susan deserves a special "Thank You" for giving me the space in our relationship that helped tremendously and allowed me to complete this task. And finally, my deepest gratitude goes to my mother, who supported my book idea and never gave up on me, and to my father, who always believed in me, but sadly passed on much too early in life and so could not hold this finished book in his hands.

PREFACE

The Digital Dictionary contains a collection of information that is public knowledge. However, it represents more than nine years of research time and a substantial creative compilation and editing effort that simply cannot be measured in time increments. The entire book content was assembled, cross-checked, and verified with help of the World Wide Web and various public libraries and was reaffirmed in interviews with countless individuals. The Digital Dictionary was then edited and simplified to match laypeople's terms and to make it "jargon free." It does not contain a translation for every possible technical word or phrase that is used in our vastly expanding digital world; rather, it is a compilation of definitions for the most common terms and acronyms almost every one of us uses on a daily basis. Its intended audience is not for experts in the industry... instead, it is meant to be a quick-resource guide and information-packed manual that can serve as a reference for individuals who may be new to the field and for enthusiasts who are just curious and want to know more. I also welcome any corrections, suggestions and rave reviews.

This is the first edition of the Digital Dictionary and inevitably updated Dictionaries will follow with revisions and added terminology.

NUMERICAL ABBREVIATIONS

16mm - Is an industry abbreviation for sixteen millimeter motion picture film. It was introduced in 1923 when George Eastman introduced a safe, non-flammable and affordable film for the amateur movie maker.

16:9 widescreen - aka 16:9 aspect ratio. It describes a format for televisions in which DVD video is the first format to support and produce pictures with full vertical resolution on 16:9 widescreen TV sets. In the past, widescreen pictures would have been letterboxed within a 4:3 video frame and then magnified by the 16:9 TV sets, resulting in reduced vertical resolution.

1080i - aka 720p. The new standard which determines the types of high definition signals that a TV is capable of displaying. An HDTV monitor can accept signals in either 480p or 1080i format, with some models up-converting 480p to 540p for display.

1280 x 720 Pixel Resolution - refers to the high pixel count of multimedia displays, which allows them to accept and display native 720p HDTV program material, providing incredibly detailed, high-resolution images.

2:3 pulldown - A method for converting 24fps film to 30fps video.

2:3 pullup - A method for converting 30fps video to 24fps film.

2D graphics - Computer graphics that do not use any 3D techniques and carry no essential depth information.

24p - a term for 24 prints or frames in film technology. The latest camcorder models are able to record in 24 frames to imitate the film feeling. *Author's note*: according to my own observation and that by Industry professionals, the digital image cannot quite reach the film depth and the essence a film has.

3 chip - is a technology in video cameras that refers to a separate image device, called a chip, for each of the three primary colors: red, green and blue.

3D graphics - Computer graphics that are created by using three-dimensional models within the computer.

3G or 3g - Third Generation. Meant is the next generation of wireless technology that offers increased capacity and high-speed data applications up to 2 megabits.

3:2 pulldown - same as 2:3 pulldown.

3:2 pullup - same as 2:3 pullup.

3-Way Calling - Is a feature in the communication world that allows you to conduct a conference call between three or more different callers at the same time.

3x, 5x, 10x, 20x - Refers to the focal length ratio of a zoom lens - this is an optical, not a digital zoom.

324H - A standard relating to videoconferencing via web cams and/or set top videophones. Recipient and sender have to have the same standard to be compatible.

480p - a resolution definition for progressive scan DVD players

8mm - 1) a short term that usually refers to the width (in millimeters) of processed or unprocessed film material in either regular or super 8 for the home movie maker during 1960-1980. 2) A video tape material designed by SONY, housed in a plastic shell to be used in handheld camcorders.

802.11 - A standard for wireless computer networking technology.

A

AA - This abbreviation refers to the most common power source, the AA-size battery.

AAP - Association of American Publishers. The members of this particular Industry group represent major publishers of print materials, as well as associated professionals.

A-B roll - An editing process in which videotapes are played back by two different tape machines. As the tapes are played, portions of each one are dubbed onto a third tape, the composite master or "C" tape. Video editors use this technique to import footage from multiple source tapes. The resulting product is often enhanced by placing transitions between segments from different sources and by using special effects. The A-B roll technique has been widely replaced by a non-linear editing process.

AbEnd - Abnormal End. An error message, notifying the user of the unexpected termination of a signal that may be either recoverable or non-recoverable.

ABI - Application Binary Interface. The capability to imitate data. Based on a hardware-software platform other than the original one, which allows foreign applications to run. An example would be running a Windows application in the Macintosh operating system.

ABIOS - Advanced Basic Input/Output System. It refers to computer service routines built into IBM-compatible PC's that support multi-tasking.

ablation - Is the term for burning information into thin metal film, using a laser beam, for data storage. The process is used to create CDs.

abort - To cancel a computing procedure or phone call while in progress.

ABR - Automatic Baud Rate detection. Performed by a receiving device, this process determines the code level, speed, and stop bits of incoming data by reading a predetermined initial character.

ABS - Alternative Billing Service. Intelligent network service enabling subscribers to charge a call or Internet provided service to a telephone number, other than the one they are using, usually backed up by a credit card or ID number.

absolute time - The time that elapses from the exact beginning of an audio CD or a digital audio tape (DAT) to any given point in the program material that follows. This information is used to determine the precise start and stop times of a recorded program.

abstract class - In object-oriented programming, a class without instances. Instead, it has concrete classes as subclasses. For instance computer is an abstract class, and Macintosh is a concrete class.

AC - <u>A</u>lternating <u>C</u>urrent. In the United States, it is 120-volt electricity delivered by utility companies. It changes polarity from positive to negative 60 times per second, also known as Hertz. In other countries, the rate of changing polarity is often 50 cycles per second (cps or Hz) and the Voltage can vary up to 240 or 260 Volts.

AC Adapter - An accessory device that allows you to power your equipment from a wall outlet.

accelerator card - A circuit board that increases the processing speed and performance of a Computer's CPU. The board gets inserted into a slot on the motherboard or a slot in the housing of a laptop.

A C D - <u>A</u>utomatic <u>C</u>all <u>D</u>istribution. A method of routing and tracking telephone signals.

ACI - <u>A</u>djacent <u>C</u>hannel <u>I</u>nterference. Refers to noise, caused when two or more channels are present at frequencies that are too near at one another in the spectrum.

AC Power - A term that describes running your equipment off the wall outlet by means of an AC adapter, rather than by battery power.

acronym - Is usually an identifier derived from the letters or initials of a phrase, which is then used as an abbreviation.

AC3 - A digital audio compression system that uses auditory masking for compression. It works with from

1 to 5.1 channels of audio and can carry Dolby Surround coded two-channel signals.

active keypad cover - that, when opened, will answer a phone call. Closing the keypad cover will end the phone call. The keypad optimizes the compact pocket form for cellular phones.

active matrix display - A liquid crystal display panel that has three transistors (red, green, and blue) for each pixel

active picture - is the area of a TV frame that carries picture information. Outside the active area there are line and field blanking which roughly, but not exactly, correspond to the areas defined for the original 525 and 625 line analog systems.

active region - The portion of the video signal that is used for actual image information, as opposed to blanking, closed-captioning and time code.

active video lines - Video traces that are scanned on the screen between the horizontal and vertical blanking intervals. About 483 lines are visible or active in the NTSC system used in the United States.

actuator - The mechanism that moves the read and write heads of a disk drive across the platter surfaces.

AC wall charger - An accessory device that allows you to power and/or charge your equipment from a wall outlet.

AD converter - Analog-to-digital converter. A circuit or device that converts changes in voltage, pressure, or motion over time into a series of numbers that a computer can understand, store and manipulate.

address - refers to the use on the world wide web. Two types of addresses share common use within the Internet. The e-mail and the IP address.

add-on lens - Some camera or camcorder lenses have threads on the front edge that allow you to mount an auxiliary wide angle or telephoto lens on top of the standard lens.

AE - Automatic Exposure. Three common systems are used to automatically set the proper exposure according to the existing light conditions.
additional information
a)   programmed: where the camera picks the best shutter speed and aperture automatically
b)   aperture priority: the user chooses an aperture value and the shutter speed is automatically determined by lighting conditions
c)   shutter priority: the user chooses a shutter speed and the aperture is automatically determined by lighting conditions

AE lock - The ability to hold the current exposure settings and allow you to point the camera elsewhere before capturing the image.

AES - Audio Engineering Society. They defined a standard for Digital Audio which permits a variety of sampling frequencies, for example CDs at 44.1 kHz, or digital VTRs at 48 kHz. 48 kHz is widely used in broadcast TV production.

affine - Any linear geometric transformation including pan, rotate, scale, and shear.

ALF - Auto Lock Follow. An image tracking system. It can follow a number of points in a clip to sub-pixel accuracy and the data it produces can be used in numerous applications from stabilizing a shot to substituting images in moving video.

AF - Auto Focus. A system that automatically focuses the camera lens in relation to the image to be captured.

AGC - Automatic Gain Control. An electronic circuit that automatically controls audio or video input levels during the recording process.

AGP - Accelerated Graphics Port. An interface specification based on the PCI bus. It was designed to facilitate three-dimensional graphics. AGP creates its own dedicated channel, so the graphics controller can directly access main memory.

AGPS - Assisted Global Positioning System is a method used for determining a mobile station location in terms of universal latitude and longitude coordinates. This technology has been implemented by wireless signal providers in the United States, so emergency callers can easily be located.

AIFF - <u>A</u>udio <u>I</u>nterchange <u>F</u>ile <u>F</u>ormat. A standard file format for storing audio data.

airtime - The time elapsed between the start of a cellular phone call and the termination of such. Different providers will charge different rates for time including signal reception prior to voice transmission, such as busy signals and ringing.

aka - <u>a</u>lso <u>k</u>nown <u>a</u>s. An acronym that describes a name, an item or phrase with a second common meaning.

alarm clock - most cellular phones have an option in the settings menu that allows the user to use the phone as an alarm which can be set for a specific time and date or can be used as a daily alarm.

algorithm - A mathematical routine that solves a problem or equation.

alias - A name, usually short and easy to remember, that is translated into another name, usually long and difficult to remember.

alias effects - A term that describes contours and jagged edges usually caused by lack of resolution in a raster image.

aliens - Another term for alias effects. Some can be avoided by careful filtering or by dynamic rounding.

alkaline - A battery technology used in personal cellular telephones, in digital cameras and in today's camcorders.

alphabetic directory - Allows storage of names and phone numbers in alphabetic order for easy retrieval.

alpha channel - The portion of a four-channel image that is used to store transparency information.

alphanumeric directory - Allows storage of names and phone numbers in numerical order for easy retrieval.

ALT - 1) The alt key on a computer keyboard. 2) abbreviation for alternative.

ampere - a measurement of electrical current equal to one Volt applied to one Ohm of resistance.

AMPS - 1) Advanced Mobile Phone Service. It is used in North and South America and is also the most common system in the Asia/Pacific region and can be found in countries such as Australia, Hong Kong, New Zealand, South, S Korea ingapore, Taiwan, Thailand and Israel. 2) plural for ampere.

analog - A circuit or device having an output that is proportional to the input.

anamorphic lens - A lens that changes the width-to-height relationship of the original image. The most common anamorphic camera lenses in film work compress the horizontal axis by 50%.

animation - Moving imagery that is created on a frame-by-frame basis which can be accomplished via the use of computers.

animator - A person responsible for producing animations.

anonymous FTP - Is a hidden ID file transfer protocol which allows a user to retrieve documents, files, programs and other data from the Internet without having to establish a userID and password.

ANSI - American National Standards Institute. This organization is responsible for approving U.S standards in computers and communications.

answering machine - allows callers to leave a personal message, which then can be played back by the recipient.

anti-aliasing - The process of reducing jagged outlines of letters by smoothing out edges where individual pixels are visible.

API - Application Programming Interface. The interface by which an application program accesses the operating system and other services.

aperture - The adjustable lens opening known as iris inside the lens.

aperture grille - The masking screen inside conventional TV tubes through which electrons are passed to hit correctly colored phosphors.

aperture priority AE - Automatic Exposure is calculated based on the aperture value chosen by the photographer. This allows for depth of field control: large aperture = shallow depth of field; small aperture = deep depth of field.

APLS - Alternate Phone Line Service. This kind of phone line allows you to have two phone numbers to keep business and personal calls separate.

application - A computer program that performs a function directly for a user.

archive - A collection of data placed in long-term storage.

artifact - aka artifacting. It is misinterpreted information from a JPEG or compressed image, showing color faults or line distortion that has a negative impact on the image.

artificial intelligence sound - a circuit inside the TV which controls the sound level to prevent wide variations in volume among different TV channels and types of programming.

ASCII - American Standard Code for Information Interchange is a standard character-to-number encoding protocol, widely used in the computer industry.

assembly editing - A method of electronically marking beginning and ending scenes from a video program and re-arranging them into a new sequence.

ASN/1 - _A_bstract _S_yntax _N_otation _O_ne. A Language used to encode Simple Network Management Protocol packets.

aspect ratio - The relation of horizontal to vertical dimensions of an image. A 35mm slide frame is 3:2, a regular TV has an aspect ratio of 4:3, the HDTV technology 16:9 and a 4X5 film's aspect ratio is 5:4. In plasma and LCD screen technology, the term "aspect ratio" describes the relationship between the width and height of a TV or display screen. Traditional displays have an aspect ratio of 4:3, meaning that for every 4 inches of width, there are 3 inches of height. This results in a screen that looks almost like a square. Many newer displays have an aspect ratio of 16:9, so their screens are almost twice as wide as they are tall. Also referred to as "widescreen," 16:9 displays are better suited to widescreen DVDs and HDTV broadcasts. The term "aspect ratio" can also be used to describe the dimensions in which a film or TV program are shot. Traditional broadcasts are usually in 4:3 format, while HDTV broadcasts have an aspect ratio of 16:9.

aspherical lens - A camera or camcorder lens whose edges have been flattened so that it is not a perfect sphere, produces a superior image. Technically, an aspherical lens is a lens whose curved surface does not conform to the shape of a sphere. Lenses are usually ground or molded with spherical surfaces; because a spherical surface lens has difficulty in correcting distortion in ultra wide angle lenses or in large-aperture lenses, an aspherical lens is used.

ATA - 1) Advanced Technology Attachment. 2) the abbreviation for a Western Digital IDE disk interface

atmosphere - A depth cue that causes objects to decrease in contrast as they move into the distance.

attenuator - a device that reduces the strength of any electronic signal.

ATV - Advanced Television. The term used in the US to describe television with capabilities beyond those of analog NTSC. It is generally taken to include digital and high definition television.

audio - a general term for the sound portion of a signal.

audio head - an magnetic recording assembly that records or plays back audio signals.

audio-in/out - a jack that either delivers the audio signal into a cable or that permits the signal input into a unit for recording.

audible keypad tone - A tone confirms that a key has been pressed properly.

audio mixer - a circuit inside an equipment housing that permits several different signals to blend with each other and to create a composite track by adjusting the signal strength individually.

AUP - Acceptable Use Policy. An Internet Service Provider's statement of permitted usage.

authentication capable - If your cellular phone is equipped with an A-Key, the key will act as a PIN number to offer you and the cellular service provider an additional level of security against cellular fraud.

automatic answer - Allows a phone call to be answered without pressing the SND button. After a brief ring, simply lift the unit to your ear and begin the conversation.

automatic display timer - automatically displays the call duration at the end of each phone call.

automatic exposure - The camera automatically adjusts the aperture or shutter speed or both for the proper exposure.

automatic hands free - When this cell phone feature is activated, simply place the unit anywhere to begin a hands free conversation.

automatic lock - When activated, a cellular phone will automatically lock each time it is turned off to help prevent unauthorized use.

autofocus - The camera or camcorder lens focuses automatically at the first object in line of sight.

A/UX - Apple Unix. A version of Unix that runs on Apple computers in a network environment.

AUX - auxiliary. A designated input only circuit on audio and video equipment.

AVI - Audio Video Interleave. Meant is a kind of movie playback format for PC's running on a Windows Operating System.

A W B - Automatic White Balance. A system for automatically setting the white balance in today's digital cameras.

B

b&w - Abbreviation used to describe black and white

backbone - The top level in a network that connects other computer networks.

back channel - Return connection in a two-way data circuit, such as a coaxial cable or satellite circuit

backdoor - A breach that designers or programmers purposely leave in the security of a computer system. Some operating systems are shipped with privileged accounts, or backdoors, intended for use by field service technicians or maintenance programmers. Also referred to as a trapdoor or wormhole.

back-end – A software that performs the last stage of a process, executing a task that is transparent to the user. The term refers to network applications that run on a server without making the client aware of their operations.

background - 1) The area of a screen or frame over which images or objects are placed. 2) The most distant element in composite layering. 3) The place where less critical events or operations are conducted during shared processing in a multitasking environment. Print spooling while a document is being edited is an example, as is the ability to receive a fax while performing other computing functions.

back lit - Once activated in a camcorder, a special circuit electronically illuminates a subject from behind for even exposure.

backlight - The illumination for a color LCD display. Backlit illumination results in lighting up a subject from behind in a video production or film shoot, creating a sense of depth by separating the foreground from the background.

backup - To copy existing data onto a storage medium.

bad sector - An area on the surface of a disk that is unable to hold data reliably due to damaged formatting or a flaw in the medium itself.

balanced line - A grounded line with two conductors that carry equal voltage opposite in polarity. In a balanced-to-ground line, the impedance-to-ground levels in both conductors are equal in strength. Audio connections made with this type of line show less interference and radio frequency noise.

band - Frequency range between two defined limits. The audio band of frequencies that can be detected by the human ear lies between approximately 20 Hz and 20 kHz.

banding - An artifact of color gradation in computer imaging.

bandpass filter - A device that blocks all frequencies that are not in a specified range. It can be used to process audio signals.

bandwidth - The difference between the highest and lowest frequencies of a transmission channel, measured in Hertz.

banner - 1) Any title page automatically added to a print job by a print spooler. 2) In relation to the World Wide Web, this is an advertisement in a box, that appears on a web page and is linked to another website. It typically contains a logo and a marketing slogan intended to attract visitors.

bar code - An assortment of parallel lines. The variable thickness and separation is actually a coded number or message that an optical scanner or bar code reader can capture and that a computer can then interpret.

barker channel - A cable TV channel dedicated to promote pay-per-view events, most often by displaying crawling text.

barn doors - A set of folding flaps that cover the front of a video light. They are adjustable to control the illumination of a subject.

barrel distortion - A common geometric lens distortion causing an image to drop toward the center or curve outward.

baseband - A method of transmitting digital information over short distances. The cable's complete bandwidth is used to transmit a single signal.

base station - 1) A site for cell phone antennas. 2) a wireless transmitter developed by Apple Computers, based on the 802.11 protocol, aka AirPort.

battery meter - A visual indicator showing the estimated charge left inside a battery cell.

BASIC - Beginners' All-Purpose Symbolic Instruction Code. A computer programming language that is made up of logic statements and other English commands as well as mathematical formulas.

baud - A measurement that reflects the speed at which a modem communicates.

BBS - 1) British Broadcast System 2) Bulletin Board System. It is an electronic messaging services which archives files and any other messages of interest to the bulletin board system's operator.

BCNU - short for be seein' you

best boy - The chief assistant to the gaffer on a movie or television set.

Beta - The testing stage of a product or service.

BetaCam - aka BetaSP, is a half inch, high speed video format mainly used for broadcasting, video archives and in TV station applications.

Betacam SX - A digital tape recording format developed by Sony which uses a constrained version of MPEG-2 compression using video tape at 11/42-inch width.

BetaMax - A half-inch video format for consumers introduced by SONY in 1974.

BG - background. The part of an image that fills the screen behind an object in the foreground.

binary - Two digit numeral system. The two numbers used are '0' and '1'.

binary file - Any file that is not plain ASCII text. For example: executable files, graphic files and compressed files.

bit - is a b(inary) (dig)it. It is the smallest unit of memory. Binary digits are 0 and 1, also known as on's and off's.

bit depth - This refers to the color or gray scale of an individual pixel. A pixel with 8 bits per color gives a 24 bit image. You can arrive at this by multiplying the bits (8) by the three primary colors (3): 8x3 = 24. The 24 bit color image resolution is 16.7 million colors.

bitmap - The method of storing information that maps an image pixel, bit by bit. There are many bitmapped file formats, .bmp, .pcx, .pict, tiff, .tif, .gif, and so on. Most image files are bit mapped.

black box - usually refers to a device for televisions that allows a scrambled premium cable signal to be descrambled and be watched without paying a fee. The use of devices like this is unlawful and when in use or when detected, will be prosecuted by the signal provider.

black level - the portion of a video signal that determines pure black in the image.

bleed - Is a Printing term, referring to an image or area that extends to the edge of the printed piece.

blooming - A visual effect caused by overexposing a CCD to too much light, This "digital overexposure" can cause distortions of the subject and the images' color information.

blue laser DVD - A DVD technology to match high definition television devices, such as TV systems with 1080i or 720p capability.

bluescreen - Is used as a term to describe a special effect in film and video. An object is placed in front of a blue background and then recorded. During playback the background is then replaced by life video or previously recorded footage. The color blue has lately been replaced by the color green to match more creative options.

bluetooth - A wireless standard for connecting cameras, PDAs, laptops, computers and cell phones when in range…usually around 30 feet or 10 meters. Bluetooth technology uses very high frequency radio waves and easily establishes a connection. This technology allows you to use a wide range of computing and telecommunications devices.

BMG - Bit mapped graphic. It is a file format popular with Windows computers. It is an uncompressed file format like TIFF.

bounce - aka undeliverable. 1) The return of a piece of mail because of an error in the delivery process. 2) the reflection of a light beam off a reflector, a wall or ceiling.

BPS - Bits per second. A measure of how fast the number of 0s and 1s travel through a channel per second.

bracketing - Bracketing is a technique used to take a series of images of the same scene at a variety of different exposures that "bracket" the metered or manual exposure.

brightness - The value of a pixel in an electronic image, representing its lightness value from black to white. Usually defined as brightness levels ranging in value from zero = black to 255 =white.

broadband - Is the term for high-speed transmission. The term is commonly used to refer to communications lines or services at T1 rates (1.544 Mbps) and above.

browse - 1) to shuttle back and forth on a video timeline to view clips for editing. 2) a term used on the Internet when jumping from website to website.

browser - the upper, highlighted portion of a website. It is program that accesses and displays files and other data available on the Internet and other networks.

BTW - By The Way. Commonly used during computer chat.

bug - An error in a computer program that causes the system to behave erratically, incorrectly or to stop altogether.

buffer - A temporary storage area usually held in the RAM portion of your computer. The purpose of a buffer is to act as a temporary holding area for data that will allow the CPU to manipulate the data before transferring it to a device, such as a printer.

built-in mini-browser - A feature on your cell phone. Your signal provider may offer you this feature which allows you to use the display screen and read e-mails or view web content. Because of it's small size it is commonly referred to as a mini-browser.

burn in - Double exposure of an element over a previously exposed piece of film.

burning - In the computer and media Industry this term refers to the writing process of data to a CD or DVD.

burst mode - The ability to rapidly capture images as long as the shutter button is held down.

bus - one complete channel in a video or audio system. Frequently used in special effect mixers and switchers as A-Bus, B-Bus and so on.

byte - a computing term which equals eight bits of memory.

C

cable modem - A device that enables you to connect your PC to a local cable TV line and receive high speed data from the world wide web.

calibration - Adjusting the color of one device relative to another.

call in absence indicator - A cellular phone feature. If the unit is left active and an incoming call is not answered, a message indicator will be displayed or an audible signal can be heard to inform the user of a call attempt.

call alert - An audible tone is send to your cellular phone to alert you of an incoming call.

call barring - Allows you to set your phone to prohibit certain incoming or outgoing phone calls, if this feature is provided by your network.

caller ID - If subscribed to this feature and having a caller ID box or display window on the phone itself, the calling party's name and number are displayed.

call forwarding - Allows you to forward all calls to an alternate phone number.
call timer - Allows tracking of airtime usage to monitor phone expenses.

call waiting - While making a call, the phone will alert you that another person is trying to call you. You can choose to answer the call and putting the present caller on "hold" by pressing the "flash" button.

can - aka canister. This term refers to the storage container for film reels.

card reader - A device that accepts flash memory cards to transfer the data to the computer.

cardioid - Refers to the heart shaped recording pattern of a microphone.

CCD - Charged Coupled Device, is a light sensitive chip used for image gathering. In their normal condition these are grey scale devices. To create color a color pattern is laid down on the sensor pixels, using a RGBG color mask (red, green, blue, and green) The extra green is used to create contrast in the image. The CCD Pixels gather the color from the light and pass it to the shift register for storage. CCD's are analog sensors, the digitizing happens when the electrons are passed through the A/D converter.

CCSM - Color coded status messages in cell phone technology. Lights indicate: In Use, Roam and No Service.

CCU - Camera Control Unit. A box, usually wired between a camera and a switcher, that has adjustments for all camera functions like iris, chrominance, luminance and focus.

CD - Is a Compact Disc, used to store compressed video, data and music files.

CD drive - Compact Disk drive. It is a device found in desktop computers or laptops to play back CD-ROMS or CD's in general including music CD's. A CD-Drive cannot play back DVDs.

CDMA - Code Division Multiple Access. It separates communications by code. The voice is broken into digitized bits, and groups of bits are tagged with a code. Groups of bits from one call are randomly transmitted along with those of other calls. Then they are reassembled in the correct order to complete the conversation.

CD-R - Compact Disc recordable. A disk on which data, music or video files can be recorded on once.

CD-RW - Compact Disc Re Writeable - a kind of disk that can be erased and new content can be re-recorded many times.

CDTV - Conventional Definition Television. The analog NTSC, PAL, SECAM television system with normal 4:3 aspect ratio.

cell site - The local cellular tower and radio antenna, including the radios, controller, switch interconnect, etc., that handles communication for subscribers in a particular area.

cellular carrier - One of the two competing companies in a given geographic area that operate a cellular phone system.

cellular service - aka. cellular network. Is a communications service that provides two-way voice and data communications through handheld, portable, and car-mounted phones or through wireless modems and other devices such as laptop computers and electronic notebooks. Geographic coverage areas for cellular service are very large and can cover cities, counties, and even entire states.

cellular service provider - A company affiliated with a cellular carrier that provides cellular service to their customers.

celluloid - aka film. Highly flammable substance made from cellulose nitrate and camphor; used in the motion-picture industry. Still widely used by professionals, film schools and during 1930 to 1980 by home movie enthusiasts.

center weighted - A term used to describe an auto exposure system that uses the center portion of the image to adjust the overall exposure value.

CFC - Compact Flash Card. It refers to a storage device in form of a small and thin rectangular plate, that can be inserted in cameras, camcorders and PDA's to hold data.

CGI - Computer Generated Imagery. An image or images created or manipulated with the aid of a computer. Often used to refer specifically to computer generated titles.

channel - One piece of information stored within an image. True color images, for instance, have three channels: red, green and blue.

chat - A protocol which allows two or more people on different computers to communicate in real-time.

chip - aka wafer. A device on which one or more circuits are etched or photographically printed on in layers, connecting active and passive devices within the units' structure.

chroma - The color intensity of an image element. The chroma is made up of saturation and hue values. It is a separate value from the luminance value.

chroma key - A technique in film and video production that allows to separate an object from its background based on colors that are unique to either the foreground or background.

chromatic aberration - Also known as the "purple fringe effect." It is common in two megapixel and higher resolution digital cameras where a dark area is surrounded by a highlight. Along the edge between dark and light a line of purple or violet colored pixels appears.

chrominance - The color portion of a video signal, carrying the hue and saturation values.
CIFF - Camera Image File Format is an agreed method of a digital cameras' image storage.

cigarette lighter adapter - An accessory device that allows you to power and/or charge your phone or laptop battery from a car's cigarette lighter.

cinemascope - A film format that produces an image with an aspect ratio of 2.35:1. Cinemascope was originally a specific process developed by 20th Century Fox in the 1950s, but has become a generic term for the film format used with Panavision lenses.

client - A computer system or process that requests a service from another computer system or process. A workstation requesting the contents of a file from a file server is a client of the file server.

clip - A short segment of film or video footage, often cut from the raw footage, which can be used as a reference for color and lighting.

CMOS - Complementary Metal Oxide Semiconductor. An alternative imaging system used by digital cameras.

C M S - Color Management System. A software program or a software and hardware combination, designed to ensure color matching and calibration between video or computer monitors and any form of hard copy output.

C M Y K - Cyan, Magenta, Yellow, Black; these are printer colors used to create color prints. When used by a printer the CMYK is also known as a reflective color since it is printed on paper or reflective films.

coax - refers to a single conductor cable that is used for the transmission of audio and video signals. A coax cable is also known as 75 Ohm cable.

codec - A MPEG compression format for video.

color balance - The accuracy in which the captured colors match the color scheme in the original scene or image.

color cast - An unwanted tint of one color in an image caused by a disproportionate amount of cyan, magenta, and yellow.

color copier - Color printing device using electrostatic and CMYK pigments.

color correction - The process of correcting or enhancing the color of an image before it is printed.

color depth - Computer displays use 24-bit true color technology. The pixel depth displays 16 million colors, about the same number as the human eye can recognize.

component video - Is a video signal in which the luminance and chrominance elements are maintained separately.

composite video - Is a video signal in which the luminance and chrominance elements are combined into a single signal.

COM port - Communication port. Todays computers may have more than just one communication port: serial, USB1 or USB2 or even firewire 400 or 800, all designed to communicate and transfer data between computer and external peripherals like printer, scanner, hard drive and camera or camcorder.

compression - Is the virtual shrinking of information. Digital images create extremely large files. A low-resolution 640x480 image has 307,200 pixels. If each pixel uses 24 bits (3 bytes) for true color, a single image takes up about a megabyte of storage space. To make image files smaller almost every digital camera uses some form of compression.

continuous autofocus - provides a crisp and in focus image to a camera or camcorder, even before the shutter release is pressed or the recording has started.

contrast - A measure of rate of change of brightness in an image.

control track - A signal, recorded onto video tape as frame reference. It is a magnetic equivalent of sprocket holes in film. One of the main purposes of "striping" tapes with a control track, is to have a reference to add pictures and audio later at a specific point, as in insert editing. Control tracks are not used in disk recording or non-linear editing.

copyguard - A patented code, imbedded in the video track, that when activated, prevents the copying of a video and audio signal. Activation occurs when cables are hooked up to the output ports of the playback equipment.

cordless - A term generally applied to radio technology in which a telephone handset is used within a restricted distance from its corresponding base station.

CPU - central processing unit.

CPS - cycles per second, also known as Hertz.

cracker - A cracker is an individual who attempts to access computer systems without authorization.

cropping - The intentional or unintentional removal of part of an image that is outside a specific boundary.

CRT - cathode ray tube in televisions or monitors.

cue - aka. cueing up. 1) To advance audio or video tape to a specific part or spot in its video or audio footage. 2) An electronic marker, placed on a video or audio tape, to indicate an editing point or the beginning of a recording.

cursor - A graphical marker, usually controlled by a device such as a mouse or a tablet, that is used to point to a position or object and is visible on a computer monitor.

cut - the intentional ending of a scene or sequence in film or video.

cyberspace - A term used by William Gibson in his 1984 science fiction novel Neuromancer in which he describes the interconnected world of computers and society.

D

D1 - A format for digital video tape recording in a 4:2:2 standard using 8-bit sampling. The video tape is 19 mm wide and allows up to 94 minutes of recording.

D2 - The standard for digital composite PAL or NTSC signals recorded to tape. Used is a 19 mm tape and the recording time is up to 208 minutes on a single cassette.

D3 - A VTR standard using 11/42-inch tape for recording digitized composite PAL or NTSC signals sampled at 8 bits. Cassettes are available for 50 to 245 minutes.

D5 - A VTR format using the same cassette as D3 but recording component signals sampled at 10-bit resolution. With internal decoding the D5 VTR can play back D3 tapes and provide component outputs.

D6 - A digital tape format which uses a 19mm helical-scan video tape to record uncompressed High Definition Television material at 1.88 GB per second.

D16 - A recording format for digital film images making use of standard D1 recorders.

dark fiber - is an optical fiber infrastructure that is in place but is not being used. In an active optical fiber line, all information will flow in form of light pulses, so the "dark" means no light pulses are being sent or received.

DAT - Digital Audio Tape. It allows recording of instruments and vocals in a digital format onto a 4mm wide tape. Usually used by musicians to do a raw cut of their work or to archive the final mix.

data service - Is the electronic transfer of data or digital information.

dB - decibel. A unit of loudness, measured on a logarithmic scale. The human ear can perceive a 1 dB change in loudness.

DC - 1) Direct Current; refers to battery power as in a 9 volt battery. 2) An abbreviation for digital camera.

decoding - The restoration of encoded data to its original form.

decoder - A device that separates a composite video signal into a component video signal.

decompression - The process by which the full data content of a compressed file is restored.

decryption - is the process of converting encrypted data back into its original form, so it can be understood.

defragment - To reorganize the information on a storage device or a computer file to eliminate fragmentation.

deinterlace - The process of splitting two fields that make up a video image into two separate images.

densitometer - A tool used to measure the amount of light that is reflected or transmitted by an object.

depth of field - aka range of focus. Controlled by the aperture value of the lens and usually achieved by having the focus sharp in the foreground while the background is out of focus.

dialer - A program which establishes and maintains your connection to the Internet.

dialup - A temporary connection between a computer and the world wide web, established with a modem over a standard phone line.

diffusion - A method of decreasing the light intensity in film projectors. Accomplished by letting the light beam pass through a heat resistant and milky tinted glass.

diffusion dithering - A method of dithering that randomly distributes pixels instead of using a set pattern.

DigiCam - A camera that is capable of capturing and storing an image or a series of images on a digital medium such as a chip, a flash or smart memory card or even video tape.

digital - The general term for communications and media technology that digitizes transmissions into binary code.

digitizer - A hardware component which converts an analog signal to a digital signal.

Digital-8 - A consumer video format introduced by SONY. It is 8mm wide, records video and audio in a digital high speed fashion and is housed in a compact plastic tape shell.

digital film - Term used to describe solid state flash memory cards.

digital zoom - A fifty percent digital magnification of the center of an image. Digital zooms by nature generate less than sharp images because the new "zoomed" image has been interpolated.

digitization - The process of converting analog information into a digital code for use by a computer or other digital playback equipment.

director - The person with the responsibility for overseeing the creative aspects of a video or film production.

displays - Refers to a variety of view screens available on cell phones and computer systems.

dissolve - A specific transition effect in which one scene gradually fades out at the same time that a following scene fades in.

distinctive ring - Is simply a setting on your phone that recognizes a call from a designated list of phone numbers and if activated, the phone will ring in a special way.

dithering - A method for simulating many colors or shades of gray with only a few available ones. A limited number of same-colored pixels located close together is seen as a new color.

DIN - Deutsche Industrie Norm. The German equivalent for Industry Standard.

DMR - Digital Media Recorder aka DVR "Digital Video Recorder". This acronym refers to a stand alone recorder that has a built in hard drive on which audio and video signals can be recorded and then played back in DVD quality on a TV set. Current units are capable of recording 30+ hours of footage, but the recording time and disk space will increase as technology evolves.

DNS - Domain Name System. It is a general purpose distributed, replicated, data query service. The principal use is the lookup of host IP addresses based on host names. The style of host names now used in the Internet is called "domain name", because they are the style of names used to look up anything in the DNS. Some important domains are .COM (commercial), .NET (network), .EDU (educational), .GOV (government) and .MIL (military). Most countries also have a domain. For example, .US (United States), .UK (United Kingdom), DE for Deutschland (Germany) and .AU (Australia).

DOF - Abbreviation for depth of field.

dolly - A rolling device in the film and video industry that allows a camera to create smooth horizontal motion that follows the action.

domain - A group of computers whose hostnames share a common suffix, the "domain name".

domain name - Is an address of a computer network connection that identifies the owner of the address.

download - To transfer images or data from the camera to the computer using a cable attached to either the USB or firewire port.

DPI - Dots per inch. A measurement value used to describe either the resolution of a display screen or the output resolution of a printer.

DPOF - Digital Print Order Format. Allows you to embed printing information on your memory card. Some photo printers with card slots will use this info at print time. Mostly used by commercial photo finishers or print kiosks which can be found in malls or chain stores.

DRAM - Dynamic Random Access Memory. A type of memory that is volatile - it is lost when the power is turned off.

DRAM Buffer - All digital cameras have a certain amount of fixed memory inside to facilitate image processing before the finished picture is stored to the flash memory card.

dropout - A momentary loss of an audio or video signal, or both, during playback.

DSL - Digital Subscriber Line is a technology for bringing high-bandwidth information to homes and small businesses over ordinary copper telephone lines.

DTE - direct to edit. An editing process in which video clips can be imported to a software

DTMF - Dual Tone Multi Frequency are the tones that your phone transmits to communicate with tone activated cell phone systems like voice menu driven or bank-by-phone systems.

DTV - Digital Television. A television with a built-in module to receive and display digital signals.

dual battery capability - allows you to use a main and auxiliary battery on your phone for extended talk times.

dual mode capability - The cell phone is capable of accessing digital or analog channels to provide a better chance of obtaining cellular service.

dub - aka dubbing. The process of making a copy of video or audio material.

DV - Digital video. A technology to record audio and video signal in a digital format.

DVCAM - aka digital full size video recording format introduced by SONY.

DVCPRO - aka digital full size video recording format, introduced by Panasonic.

D-VHS - a video format by JVC, in which a half inch video tape carries a resolution of over 1000 horizontal lines.

DVD - Digital Versatile Disc is pretty much the viewing standard for recorded movies in the beginning of the new Millennium, the 21st Century.

DVD audio - A high capacity read-only optical disc. Capable of rapid data transfer that may be used for the playback of high-quality audio using 24-bit linear pulse code modulation sampled at 48 or 96 kilobytes per second.

DVD+MRW - a planned DVD format; also known as Mount Rainier drag-and-drop file access support.

DVD-R - is a write once recordable disk format which allows excellent compatibility with both standalone DVD-video players and DVD-ROM drives. There are two main types of 4.7 GB DVD-R discs: DVD-R for General Use and DVD-R for Authoring. Most consumer drives use the former cheaper General Use discs, while many higher end professional drives use Authoring discs. The correct media type appropriate for the drive must be used when burning. However, once burned, the discs should be able to be read in either drive type.

DVD-RAM - is a format primarily used as a data solution, although the type of data stored can include many types of data including video with a typical capacity of 4.7 GB single sided discs or double-sided 9.4 GB discs. These discs are traditionally housed inside a plastic shell, the so called cartridge. The Type 1 cartridge cannot be opened, but newer Type 2 cartridges can be opened and the disc can be used in drives or standalone DVD-Video players/recorders which do not use the DVD-RAM cartridge loading mechanism. DVD-RAM is a very robust data storage solution, theoretically allowing greater than 100000 rewrites per disc. The main drawback of DVD-RAM is its very limited read compatibility in DVD-ROM drives and standalone DVD-video players.

DVD-ROM - A high-capacity read-only optical disc capable of rapid data transfer that may be used as a general-purpose computer storage device. A DVD-ROM may hold any type of digital data and is readable by a DVD-ROM drive connected to a computer.

DVD-RW - uses rewritable discs which are rated at more than 1000 rewrites in ideal situations. Unfortunately, DVD-RW does not enjoy the same excellent compatibility with DVD-ROM drives and standalone DVD-Video players that DVD-R enjoys. Another drawback is that unlike DVD-RAM, one must generally erase a DVD-RW disc before reuse. Most DVD-RW drives should also be able to record to DVD-R. Specific drives can also record to CD-R and CD-RW in addition to DVD recording. For instance the Apple SuperDrive, found in high end Macintosh computers used in multimedia creation, is a DVD-R/DVD-RW and CD-R/CD-RW capable drive.

DVD-Video - A disc that is used for interactive playback of movies and games or other video, audio, and graphic content using the MPEG-2 video compression format. The DVD is played on a DVD player and viewed on a television monitor. It has become one form of a "set-top box." These discs may also be played on a computer with a special hardware decoder and supporting software.

DVD+R - is defined as a subset of DVD+RW. Using dedicated DVD+R write once discs, which will be substantially cheaper than DVD+RW discs, users can record material in the same way as with DVD+RW discs, but without the ability to erase the disc or re-record onto them. DVD+R discs can be recorded by any DVD+RW PC drive or standalone video recorder. DVD+R shares most of the characteristics of DVD+RW, such as the same storage capacity of 4.7 GB and the same usage applications, combined with a higher level of compatibility due to a higher reflectivity factor of the disc. As with DVD+RW, there will only be one format of DVD+R discs, which are usable in both video recording and data applications. DVD+RW is designed from the start to be compatible with existing DVD-ROM drives and DVD-Video players, both on a physical as well as a logical level. This means that a DVD+RW disc recorded in a DVD+RW video recorder can be played in virtually all DVD-Video players or DVD-ROM equipped PCs, and that any DVD+RW disc recorded with data on a PC DVD+RW drive can be read by most DVD-ROM drives.

DVD Video Parameter Settings:

Frame Size: 720x480 (NTSC) or 720x576 (PAL)

Frame Rate: 29.97 frames/second (NTSC) or 25 frames/second (PAL)

Video Data Rate: 4~8 Mbps CBR or VBR (Constant/Variable Bit Rate)

Audio Settings: Stereo, 48 kHz and 192~384 kbps MPEG audio

DVD drive - A device that is capable of high definition video playback with over 500 lines of resolution. Starting with the fourth generation of standalone DVD players or DVD drives in computers, the playback capabilities include MP3, CDR/RW and VideoCD with the appropriate software.

DVD player - Found inside a computer or as a stand alone device. Capable of playing back high resolution video with sound through a monitor or a Television set.

DVD recorder - A stand alone or computer recording device, capable of recording video and audio signals onto a 4.7gb or 9.4gb DVD.

DVD+RW alliance - A group of companies that co-developed and promote the DVD+RW standard: Philips, Hewlett-Packard, Sony, Yamaha, Ricoh and Mitsubishi/Verbatim. In addition to these, a couple of dozens of other companies work together in the DVD+RW Alliance to share knowledge on the creation of DVD+RW products.

DVR - Digital Video Recorder aka DMR.

dye sub - Dye Sublimation is a printing process where the ink is thermally transferred to printing media. Usually expensive but it yields prints that rival real, wet-processed photographs.

dynamic range - A measurement of the accuracy of an image in color or gray level. More bits of dynamic range results in finer gradations being preserved.

E

ear to mouth ratio - The relative positions of the mouth and ear on an adult head. Motorola pays particular attention to this ergonomic factor in designing all its phones.

Eb - exabit. A measurement for data storage. It equals1,000,000,000,000,000,000 bits.

EB - exabyte. A measurement for data storage. 1,152,921,504,606,846,976 bytes or 1024 petabytes.

EIAJ - Electronic Industries Association of Japan.

electronic lock - When activated the phone will automatically lock each time it is turned off to help prevent unauthorized use.

electronic scratchpad - Is a message pad on a cellular phone and allows storage of a number during a call.

electronic viewfinder - aka EVF

Email - Stands for electronic mail and refers to written content send from one Internet user to another or shared text and picture messages between e-mail account holders.

Email address - An address that is used to send electronic mail to a specified destination. For example, "info@videopros.net" is the email address for the information department inside the videopros.net domain.

encoding - The translation of binary data into a 7-bit ASCII representation using a data encoding scheme.

encryption - is the conversion or translation of data into a form that cannot be easily understood by unauthorized people. Usually used to transmit highly sensitive messages with a great need for security.

ENG - Electronic News Gathering. A term used by news stations and freelance videographers.

enhanced progressive scan doubler - A technology that provides even more image enhancement than the progressive scan technology by improving the quality of diagonal lines in a moving picture. It displays a full frame of video in 1/60th of a second by de-interlacing the incoming video signal and progressively scanning the image. With this technology the viewer generally won't see the horizontal scan lines that can be visible in ordinary television images.

E P P - Enhanced Parallel Port. A hi-speed, bidirectional printer port on computers. Some digital cameras and scanners use the EPP port to transfer data.

EPROM - A programmable read-only memory that can be erased by exposure to ultraviolet light and then reprogrammed.

erase head - A rotating device that erases an existing signal on a video tape before recording a new image.

ESN - Electronic Serial Number. A unique unchangeable number, that is built into a cellular phone and is transmitted by the unit, identifying itself within the network.

ETTL - Canon's evaluative trough the lens exposure system that uses a brief pre-flash before the main flash to calculate the exposure index.

EV - Exposure Value. It usually means the ability to override the auto exposure system to lighten or darken an image.

EVF - Electronic View Finder. A small b&w or color LCD with a magnified lens that functions as an eye level viewfinder to aim at the to be captured image. Usually found on video camcorders or digital cameras.

EXIF - Refers to the embedded camera and exposure information that a digital camera puts in the header of the JPG files it creates.

exposure - The amount of light that reaches the image sensor of a camera or camcorder. It is controlled by a combination of the lens aperture and shutter speed.

exposure compensation - Is a term used when lighten or darken the image by overriding the exposure system.

extension - A compilation of three or four letters. Usually placed after a filename and a "dot", referring to the type of file, so that a computer program that uses the data in a file is able to recognize and possibly access data within the file. .exe = executable file type, .JPG = image type file, .mov = movie file and so on.

F

FAQ - Frequently Asked Question. In its usual context, FAQ refers to collected answers to often-asked questions for users.

FCC - Federal Communication Commission

FDD - Floppy disk drive, the most common storage medium, holding 1.44MB of data on a 3-1/2" diskette, that fits in most of today's PC's.

field - Is one half of a Television picture. Two fields, when interlaced, build a frame.

file - Is a collection of information. Such as text, data, or images, compiled and saved on a disk or hard drive.

file format - A type of program or data file, formatted in a language that your computer software understands.

file server - A storage system that provides data files to all connected users of a network.

file transfer - The copying of a file from one computer to another over a computer network.

film transfer - a technique to preserve film material to a new medium. With the use of a telecine or wet gate unit it allows film material to be captured by a high resolution camera and then recorded directly onto video tape, DVD or computer HD for further manipulation.

firewire - Also known as "iLink" and officially designated as the IEEE1394 protocol. It is a high-speed data interface on digital camcorders, digital still cameras and other equipment, allowing the electronics to communicate with the computer.

firmware - An often used program or instruction set stored on a ROM chip. Usually refers to the ROM-based software that controls a unit. Firmware is found in all computer based products from Cameras to Digital Peripherals. Manufacturers of DVD burners and DVD players issue frequent firmware updates for their units to match the current recording speed for blank media.

fixed aperture - Normally when a zoom lens moves from wide angle to telephoto the aperture changes. If the camera has an option to fix the aperture value then it remains constant, regardless of focal length.

fixed focal length - A term that describes a non-zoom lens, it is fixed at a given focal length and does not change.

fixed focus - A lens that is preset to a given focal distance, it has no auto focus mechanism to give the camera the maximum depth of field.

flash - An external or built-in auxiliary light to supplement current or natural lighting conditions. The image usually improves in color and picture sharpness.

flash memory - This is the storage or "film" for digital cameras, it can be erased and re-used many times. It is non volatile memory and data is preserved even when it is not powered. There are several memory types for digital cameras and camcorders, usually all different in size and shape. The most popular ones are called CompactFlash, SmartMedia and MemoryStick.

flashpath - A device that resembles a 3-1/2" floppy diskette and allows a SmartMedia card or MemoryStick module to be inserted and be read in a standard 1.44MB floppy disk drive and the data is then transferred to the computer.

flat bed scanner - An optical scanner in which the original image remains stationary while the sensors passes over or under it. The scanned material is held flat and scanned using a reflective process.

flying erase head - A rotating device in video recorders that erases a signal diagonally, allowing the seamless frame to frame integration into existing video footage,

floppy disk - A paper thin magnetic data storage device, housed in a plastic shell, which can be inserted into a computer's "floppy drive" and its data can be retrieved.

fluid head - the upper part of a tripod in which the moving parts are imbedded in liquid, to insure smooth movements.

focal length - A lens' angle of view, most commonly indicated as wide angle, normal or telephoto.

focus lock - Is a certain technique of capturing an image by first pre-focusing the camera and then moving it to re-compose the image before capturing it. That style is accomplished by pressing the shutter button half way down and keeping it held at that position while moving the camera to another point before pressing it all the way to capture the image.

FPX - FlashPiX is a multi-resolution image file format.

fragmented - a term used to describe the empty spaces on a hard drive platter between data blocks, which could cause non-continuity on data or movie files due to read errors.

frame - One of the still pictures that make up a video. A frame has two fields.

frame rate - The number of frames that are shown each second in video. Live action relates to a video frame rate of 30 frames per second in the NTSC standard and 27 frames in the PAL standard.

frequency - The property or condition of reoccurrences at frequent intervals.

f-stop - A numerical designation that indicates the size of the aperture. It is inversely proportional as a smaller number like F2.8 is a large opening and a large number like F16 is a relatively small opening.

FTP - File Transfer Protocol. A protocol which allows a user to access and transfer files to and from another host over a network.

full bleed - A printing term used when an image or printed area extends to the edge of all four sides of the printed paper.

full duplex - Incoming and outgoing audio can occur simultaneously. A user can speak and listen at the same time.

function keys - Are the non-numeric keys on the upper part of a computer keyboard, marked F1 to F12 and used to access or navigate menu features, as well as performing memory functions.

FX - abbreviation for special effect in film and video.

FWIW - For What It's Worth

FYI - For Your Information.

G

gaffer - Is a construction worker, who helps building and disassembling backgrounds or sets in film and video production.

gamma - A measure of the amount of contrast found in an image according to the properties of a gradation curve. High contrast has high gamma and low contrast low gamma.

gamma correction - In reference to displaying an image accurately on a computer screen, gamma correction controls the overall brightness of an image. Images which are not properly corrected can look either overexposed or too dark.

Gb - gigabit. A measure of storage. It equals 1,073,741,824 bits

GB - gigabyte. A measure of computer memory or disk space consisting of about one billion bytes.

GMA - Glass Mount Antenna. Here we are referring to a mobile antenna type, which usually mounts on the rear window of a vehicle, without drilling any holes.

GIF - Graphic Internet File. It is a graphic file format used mainly for Web graphic or small animated files. Not recommended for photos as it only contains a maximum of 8 bit and a maximum of 256 colors.

GPS - Global Position Service. A system of satellites that is able to determine the latitude and longitude of a receiver on Earth by calculating the time difference for signals from different satellites to reach the receiver.

GPRS - General Packet Radio Service is a wireless data transmission service based on packet transmission. For example, if an e-mail is sent by GPRS it will be reduced into 'packets' of information. Each individual packet travels to its destination by the quickest possible route. This means the different packets from the same mail can travel separately through foreign networks and around the globe in order to avoid obstructions. At the pre-set destination they are rebuilt and presented to the recipient as a whole message again.

gradation - A smooth transition between black and white, one color and another, color and no color.

gray level - The brightness of a pixel. The value associated with a pixel representing it's lightness from black to white. Usually defined as a value from 0 to 255, with 0 being black and 255 being white.

gray scale - A term used to describe an image containing shades of gray as well as black and white.

grip - The person on the movie set of a film or video production who positions the camera according to the director's instructions. The camera itself may be placed on a dolly, a jib, or on any other surface that achieves the desired motion, the desired shooting angle or perspective.

GSM - Is the pan-european standard for digital cellular telephone service. It is also one of the technologies available in the US. GSM was designed for markets to provide the advantage of automatic, international roaming in multiple countries.

GUI - graphic user interface

guide number - The output power rating of a electronic flash unit.

H

hacker - A person who's delight it is, having an intimate understanding of the internal workings of a system, computers and computer networks in particular. The term is often misused in a pejorative context, where cracker would be the correct term.

HAD CCD - A CCD image device by SONY, in which the HAD stands for hole accumulated diode.

halftone image - An image reproduced through a special screen made up of dots of various sizes to simulate shades of gray in a photograph. Typically used for newspaper or magazine reproduction of images.

Handy - The widespread name for the cellular phone in Europe.

header - 1) The portion of a packet, following the actual data, containing source and destination addresses, error checking and other fields. 2) part of an e-mail message that follows the body of a message and contains information about the message originator and time stamp.

handset - A audio device with controls, held by the user on a cellular or portable phone.

HD - Hard drive aka hard disk. The internal, large capacity data storage unit in today's personal computers.

HDTV - High Definition Television. It is the new video "standard" that will display 1,125 lines of resolution.

HDML - Handheld Device Markup Language. allows Internet access from wireless devices such as handheld personal computers and phones. This language is derived from hypertext markup language.

HDMI - aka HDCP is a new standard for digital connections between A/V devices, such as a set-top box, DVD player or A/V receiver and a digital television. HDMI offers exceptional video and audio quality with a single quick-disconnect connector. It supports multi-channel digital audio transmissions and component video color spacing for true rendering of HD video. HDMI supports the HDCP copy protection standard, allowing transmission of copy-protected digital content to your display. The video portion of HDMI is also backwards-compatible with DVI-HDCP devices using a special cable adapter.

headset capable - Refers to any kind of phone, allowing the user to attach a microphone-headset accessory, to have a hands-free conversation.

Hertz - Is the assigned term for 50 or 60 cycles per second in house current.

HFO - Hands Free Operation allows you to conduct a conversation without holding the cellular phone, by usually wearing a headset or by having the phone connected to a vocal amplifier.

Hi-8 - A consumer video format in the nineties, introduced by SONY. Very popular due to its small cassette size and excellent picture quality.

histogram - A bar graph analysis tool that can be used to identify contrast and dynamic range of an image. Histograms are found in the more advanced digital cameras and most software programs used to manipulate digital images.

horizontal/vertical edge correction - A technology that sharpens the edges of objects in the displayed picture. Horizontal edge correction works in conjunction with the dynamic sharpness control and velocity modulated scan to help produce clean, crisp television images and reduce the unwanted effects of video noise. The vertical edge correction circuit works in conjunction with the digital comb filter to sharpen the horizontal edges of objects in the picture.

host - 1) A computer that allows users to communicate with other host computers on a network. 2) A person, or small group of people, who manage and monitor mailing lists, newsgroups or chat rooms. Hosts are the watchful eyes on the Internet and they are responsible for determining which contents are passed onto a list or a screen on real time connections.

hot shoe - A connector generally found on the top of the camera or camcorder, powered by the unit's internal battery, that lets you attach a light or a flash unit and trigger it in sync with the shutter.

HT - Hypertext. A link between one document and another related document on the World Wide Web. By clicking on a word or phrase that is highlighted by a different color and/or underlined on a computer screen, a user can skip directly to files related to that subject.

HTML - Hyper Text Markup Language. The coding which World Wide Web browsers read to display Web pages.

http - Hyper Text Transfer Protocol is the protocol used to transfer World Wide Web data across the Internet.

hue - A term used to describe the entire range of colors of the spectrum; hue is the component that determines what color you are using. In gradients, when you use a color model in which hue is a component, you can create rainbow effects.

I

IC - Integrated Circuit. A complex set of electronic components and their interconnections that are etched or imprinted on a chip

IDE - Integrated Drive Electronics. It is a disk drive protocol and identifies the power and data signal interface between the motherboard and the integrated disk controller and drive.

IEC - International Electrotechnical Commission. A group of engineers at the same level as ISO.

IEEE1284 - This is the high-speed bidirectional parallel port specification, used by printers and devices like card readers.

IEEE1394 - Better known as "Firewire". It is a high speed input/output bus used by digital video devices & PCs, capable of transmitting up to 400 mbps in its first generation and 800mbps in the second generation.

iLink - A SONY name for the IEEE-1394 firewire data port found on their camcorders or computers.

illuminated keypad - Found on cellular phones and new generation remote controls for equipment. It allows keypad viewing in the dark.

image processing - Meant is capturing and manipulating images in order to enhance or extract information.

image resolution - The number of pixels per unit length of image. For example, pixels per inch, pixels per millimeter, or pixels wide.

image sensor - A traditional camera exposes a piece of light-sensitive film, digital cameras use an electronic image sensor to gather the image data.

image stabilization - An optical or digital system for removing or reducing camera movement in telephoto zoom lenses.

imaging - The transmission of still images such as by faxes, pictures, or slides.

IMHO - In My Humble Opinion.

InfoLithium - Sony's "smart" lithium rechargeable battery pack. It has a chip inside that tells the camera how long it will last at the current discharge rate in minutes.

inkjet - A type of printer that sprays dots of ink onto paper to create the image.

interlaced - Term used to describe an image sensor that gathers its data by first processing the odd lines and then the even lines.

Internet - A network or collection of networks interconnected with routers. It also refers to the largest network of computers in the world.

InterNic - The Internet Network Information Center. It is a project administered by AT&T and Network Solutions and provides directory and database services for registered Internet hosts, while NSI administers the registration process.

interpolated - Software programs can enlarge image resolution beyond the actual resolution by adding extra pixels using complex mathematic calculations.

interpolated resolution - Is a method to add pixels to an image, using complex software algorithms to determine what color they should be. It is important to note that interpolation doesn't add any new information to the image - it just makes it bigger!

IP - 1) Internet Protocol. It's the network layer for the TCP/IP Protocol Suite. It is a connection switching protocol. 2) Internet provider. That is the company that allows you to tap into the internet by providing you with an IP address.

IP address - Internet Protocol Address. Location of a server assigned by your service provider.

IR - Infra Red. To humans it is an invisible beam of light to either wirelessly control a device or as a method of transferring data from camera to computer (or printer) without cables. Some cameras also employ infrared in the auto focusing system.

I S D N - Integrated Services Digital Network. It combines voice and digital network services in a single medium, making it possible to offer customers digital data services as well as voice connections through a single "wire".

ISO - is the abbreviation for International Standards Organization which has defined the light sensitivity for film. The speed or specific light-sensitivity of a camera is rated by ISO numbers such as 100, 400, etc. The higher the number, the more sensitive it is to light. As with film, the higher speeds usually induce more electronic noise so the image gets grainier.

ITAP - An application installed on wireless phones and handheld devices that allows you to type messages with just one key press per letter using the keypad. Also known as T9 TM text input.

J

J2ME™ - also known as Java™ 2 platform, micro edition or K Java™. It is a technology that allows programmers to use the Java programming language and related tools to develop programs for mobile wireless information devices such as cellular phones and personal digital assistants.

Jack - A definition for a female connector which can be found on a/v equipment to connect audio or video cables. The most common type is called RCA.

Jaggies - Slang term for the stair-stepped appearance of a curved or angled line in digital imaging. The smaller the pixels, and the greater the number, the less apparent the "jaggies".

Java™ - Is an object-oriented programming language developed by Sun Microsystems, Inc.

JFIF - A specific type of the JPG file format. Also known as EXIF.

jib - aka crane. A film or video camera is mounted on a platform at the end of a pipe like extension on top of a tripod, that allows for horizontal or vertical "moving" camera shots. The balance is established by counterweights on the opposite end of the extension pipe.

Jog - aka shuttle. It occurs when moving video footage from one frame to another, in any desired direction, during linear editing, utilizing a shuttle knob.

JPEG - Joint Photographic Experts Group. The name of the committee that designed the standard image compression algorithm. JPEG is designed for compressing either full-color or grey-scale digital images of "natural", real-world scenes.

JPG - The most common type of compressed image file format used in digital cameras. It is a "lossy" type of storage because even in its highest quality mode there is compression used to minimize its size.

JTC1 - Joint Technical Committee 1

K

Kb - kilobits. A kilobit is a unit of information, called data, equal to 1,024 bits.

KB - 1) a kilobyte, or 1024 bytes of data. 2) keyboard for a computer

Kbps - kilobits per second. One thousand bits per second. Since this abbreviation has a time value, it refers to the rate at which data is transmitted.

Kelvin - A unit of absolute temperature. One Kelvin degree is equal to one Celsius degree.

Kermit - A popular file transfer protocol developed by Columbia University. Because Kermit runs in most operating environments, it provides an easy method of file transfer.

Keypad - The buttons on the handset.

L

LAN - Local Area Network

Landscape Mode - 1) Holding the camera in its normal horizontal orientation to capture the image. 2) when selected on a printer, the image will be placed horizontally on the paper.

LNB - Last Number Backup. This function remembers the last number called, allowing for faster redialing.

LBW - Low Battery Warning! A visual and/or audible indicator, showing that the battery in your portable device is approaching discharge.

LBX - Letterbox. A TV viewing format, in which the movie is shown with a black bar on top and bottom, therefore giving the viewer a cinema-like look in his own home.

LCD - Liquid Crystal Display. This computer monitor allows easy viewing in all lighting conditions. Uses dark segments against a lighter background.

LD - Laser Disk. The term refers to a 12" first generation video disk, readable by a laser beam inside standalone players.

LED - Light Emitting Display. A monitor that provides excellent visibility in direct sunlight and in darkness.

Lithium - Some digital cameras or camcorders are packaged with a lithium rechargeable battery pack. Lithium batteries are lighter but more costly than NiMH or NiCd type of rechargeable cells. Lithium cells can be recharged regardless of their state of discharge.

Lithium Ion (Li-Ion) - A battery technology used in personal cellular telephones. Lithium Ion batteries generally provide more energy capacity than Nickel Metal Hydride batteries of the same weight.

Lock - A variety of locking options are available for portable devices like cell phones, PDA's and more.
additional information:
* Automatic Lock For enhanced security, the telephone can be programmed to automatically lock each time you turn it off.
* Display Unlock Code refers to the Unlock Code in case it's forgotten.
* Electronic Lock prevents unauthorized use of your phone. Only entry of the 3-digit Unlock Code will unlock the phone.
• Programmable Unlock Code enables the user to change the 3-digit Unlock Code should the code become compromised.

login - is an action taken by a user when he/she starts up a computer. A login may be restricted and a password may be needed to access stored files.

LOL - Laugh Out Loud or Laughing Out Loud

lossless - The digital transmission of video or audio information to another device, without loosing quality from the original.

LPCM - linear pulse code modulation

lurking - Is the non-active participation of a person in a newsgroup or "chat"-channel. An individual who is lurking is just listening to the discussion.

lumen - The unit of luminous flux in the International System of Units, that is equal to the amount of light given out through a solid angle by a source of one candlelight intensity radiating equally in all directions.

luminance - Is the intensity of light per unit area of its source

lux - is the measurement of light intensity

M

Mac - Refers to the Macintosh type of computers, developed by Apple.

macro - 1) The ability of a lens to focus very close for taking pictures of objects at a 1:1 ratio. Usually the object is as close as 3" or 75mm from the lens. 2) A specific set of instructions inside a software program to follow a certain routine or to perform a certain function.

mAh - milli Amperehour. A rating used in the consumption of power of an electronic device such as an LCD or the storage capability of a device like an NiMH or Nicad rechargeable battery.

mail server - A software program that distributes files or information in response to requests sent via email.

mailing list - An e-mail address which expands to multiple e-mail addresses. Usually they are confined to specific topics of information.

master reset - a function on computers, cell phones and PDA's. It will restore all original manufacturer's settings, but it does not clear all memory locations and call timers.

matrix - 1) The network of intersections between input and output leads in a computer, functioning as an encoder or a decoder. 2) A metal plate used for casting typefaces. 3) An electroplated impression of a phonograph record used to make duplicate records.

matrix metering - In most digital still cameras there is a matrix metering option which uses 256 areas of the frame to calculate the best overall exposure value.

MBPS - megabits per second . A measurement of data transfer in millions of bits each second.

MD - MiniDisc. As small as 3 inches in diameter it is a digital recording format like a floppy disc. The format is commonly used for audio data and has now been incorporated in several digital still cameras.

MegaBit - One million bits.

MegaByte - One million 8-bit bytes, or more precisely 1 048 576 bytes. Also known as a memory term meaning 1024 KiloBytes. Used to describe the size of a flash memory card such as 4MB, 8MB etc. When capitalizing the letters MB [megabyte] is often confused with Mb in which the first letter is uppercase and the second one lowercase [megabit] creating a completely different value: there are actually 8 bits in a byte; so 256Mb = 32MB.

Mhz - Megahertz. One million hertz or cycles per second. The acronym is used to measure radio frequency.

megapixel - Refers to the CCD resolution of one million pixels. Digital still cameras are commonly rated by

memory scroll - Allows sequential viewing of numbers and/or names stored in your cellular phone's memory, starting at a chosen point.

memory stick - A flash memory card standard from Sony. They resemble a stick of gum and come in sizes from 4MB up to 128MB.

menu - Available on computers, PDA's and cellular phones. It is a list of options from which users may choose. The options are often available through a drop-down menu bar at the top of the screen.

metering - Used to calculate the exposure from the existing light conditions.

micro browser - aka mini-browser. The software is built directly into a wireless device that allows you to access and display specially formatted Internet content, such as stock reports, news, and sports scores, using only your handset.

microdrive - A miniature hard disk drive for digital cameras and PDA devices. Packaged in a CompactFlash Type II housing and available in 170MB, 340MB, 512MB and 1GB capacities.

MIDI - Musical Instruments Digital Interface. It defines a protocol for the sound interchange between computers, musical instruments, and sound boards. MIDI is a simple serial communications bus, such as SCSI.

MIME - Multi-Purpose Internet Mail Extensions. An extension to Internet email which provides the ability to transfer non-textual data, such as graphics, audio and video files.

mini-browser - same as microbrowser

MiniCD - also known as mini disk. The term refers to the small diameter (3-inch) of CD or DVD discs with a maximum capacity of about 165MB.

mini-DV - a digital tape format for camcorder or standalone recorders. It is a 4mm wide magnetic tape, capable of recording sixty or ninety minutes of video and audio signals.

missed call indicator - The user is notified via audible tone that a call was received and not answered.

M-JPEG - Motion–Joint Photographic Expert Group. This is the proprietary extension of a JPEG compression standard for still images that adapts it for moving images. It is used to compress a stream of moving pictures at a constant frame rate, M-JPEG provides much lower compression ratios than video compression standards, such as MPEG, which capitalize on similarities between successive frames. M-JPEG files are editable but generally are not transportable to different hardware platforms for playback. M-JPEG is implemented differently depending on the hardware used to encode it, but QuickTime version 3.0 translates M-JPEG from most proprietary formats and may be used to decode common M-JPEG files.

mm - millimeter, measurement to define the focal length of a lens (i.e. 50mm)

MMC - Multi Media Card is a flash memory card used in some digital cameras and MP3 players. It is identical in size and shape to the SD flash cards.

MMP - Multi Mega Pixel. Refers to image capturing capabilities or image display capabilities in the amount of two or more mega(million) pixels in resolution.

mobile phone - Usually refers to a built into a car or a portable cellular phone.

mobile service - A personal communications service that is expected to provide two-way voice and data communications using satellites, handheld phones, and wireless modems incorporated into devices such as notebook computers. It is expected that mobile services will offer enhanced features such as call waiting and voice mail. It is anticipated that the geographic coverage will be larger than most PCS services and may be worldwide.

modem - A modem converts digital data to analog data, so that it can be sent over regular phone lines. The modem also converts data back from analog to digital.

moiré - A visible pattern that occurs when one or more halftone screens are mis-registered in a color image.

motion adaptive 3D-Y/C digital comb filter - allows a Plasma or LCD display to reproduce bright colors and fast-moving scenes with incredible clarity by minimizing the "color rainbow effect" in closely spaced patterns, compensating for the motion that occurs between fields.

MOV - Found as an extension on QuickTime MOVie files.

movie clip - A sequence of motion captured in AVI, MOV or MPEG format. Some digital cameras can capture short, about 30 seconds movie sequences, some can also record sound.

MP - Abbreviation for Mega Pixel.

MP3 - a music file format. Specifically a three layer music player format. A downloadable song from the internet is usually a compressed MP3 file. Today's technology also allows you to change a stored song on your computer into a mp3 compressed file and then burn it to a CD. Because of the compression, you'll be able to fit hours of music on one single CD, which then can be played on a DVD or CD player with mp3 playback capability.

MPEG - stands for Moving Picture Experts Group aka Motion Picture Experts Group. The name is given to a family of International Standards used for coding audio-visual information into a digital compressed format. The MPEG family of standards includes MPEG-1, MPEG-2 and MPEG-4.

MTSO - Mobile Telephone Switching Office. The computer, or switch, is the brain of a cellular system. The MTSO assigns and reassigns frequencies to each call, interconnects calls with the local and long distance landline telephone companies, compiles billing information and so on. Every cellular system has one or more MTSOs or switches.

MUD - Multi User Dungeon. It refers to adventures role playing games or simulations, played on the Internet. Devotees call them "text-based virtual reality adventures". Players interact in real time and can modify the "world" in which the game is played.

multi spot focusing - The autofocus systems uses not only one but several different portions of the image to determine the proper focus.

mute - is a control function on telephones that silences the mouthpiece inside a handset or speaker to allow private conversations without the called or calling party listening in.

N

NAM - Number Assignment Module. It is located inside transceiver and has the data about the user-- cellular phone number, lock code, timer reset code, network of choice and other operational data. The ESN is not stored here. Today's phones have an EPROM type NAM and are keypad programmable.

NAMPS - Narrowband Advanced Mobile Phone Service. NAMPS is a next generation of AMPS systems. NAMPS is a cellular service that uses digital signaling techniques to split the existing 30 kHz wideband voice channels into three 10 kHz narrowband voice channels. The result is three times more voice channel capacity than the traditional AMPS system provides. NAMPS cellular phones are manufactured for dual mode operation, and they are compatible with traditional AMPS systems.

NATAS - National Academy of Television Arts and Sciences.

NCM - Noise Canceling Microphone. A technology that screens out unwanted background noise to allow clearer conversations.

netiquette - A spin-off derived from "etiquette", referring to proper behavior on the Internet.

network - several computers and peripherals which are connected to each other and are able to communicate and share data between them.

network connection time - The time elapsed between the start of a call achieved by connecting to your service provider's network and the termination of a call achieved by pressing the end button. Network connection time includes signals received prior to voice transmission, such as busy signals and ringing.

newbie - Slang term for a user who is new to the Internet.

NIC - Network Information Center. A NIC provides information, assistance and services to network users.

NiCd - Nickel Cadmium. aka Nicad. Meant is a type of rechargeable battery. Nicad was the original type of rechargeable battery and has been pretty much replaced by the NiMH type.

night shot - a SONY technology in camcorders, allowing to record video images in low-light situations lower than 1 lux. The disadvantage: all footage will appear in a "green-tint" and has no color images.

NiMH - Nickel-Metal Hydride, a type of rechargeable battery. NiMH is the more modern type of rechargeable battery and has been touted as having no memory effect as is common with Nicad type batteries when they are charged before they have been fully discharged. NiMH may also be called NiHy by some users.

NNTP - Network News Transfer Protocol. A protocol for the distribution, retrieval and posting of articles through high-speed links available on the Internet.

noise - Pixels in your digital image that were misinterpreted. Usually occurs when you shoot a long exposure (beyond 1/2-second) or when you use the higher ISO values from 400 or above.

NSI - Network Solution incorporated

NTSC - stands for National Television System Committee, which devised the NTSC television broadcast system in 1953. This term used to describe the 60 field video output in the television standard used in the U.S. and Japan.

O

OHM - an expression for electrical resistance. For instance: if one volt of electric force is applied, then one ohm of resistance allows one ampere of current to flow.

OEM - Original Equipment Manufacturer. Usually referred to when the piece of equipment is made by one company but labeled for and sold by another company.

OLED - Organic Light Emitting Diode. This is a newly developed display technology that could replace LCD. OLED does not require a backlight like LCD displays and therefore is more energy efficient which is important to battery-operated portable devices. It also offers increased contrast and a better viewing angle which means it can be more easily viewed in bright (sunlight) conditions.

on hook call processing - aka. on hook dialing. A function that allows the user to leave the handset on its mount until the called party answers for safer operation.

one touch dialing - Used in telephones where a memory location is reserved for storing an important number. The number can be accessed and called even if the phone is locked.

optical resolution - Is an absolute number that the camera's image sensor can physically record.

optical viewfinder - An eye level viewfinder that is used to compose the photograph.

optical zoom - Means that the camera has a real multi-focal length lens, this is not the same as a "Digital Zoom" which magnifies the center portion of the picture.

orientation sensor - A special sensor in some cameras that "knows" when your turn the camera in portrait orientation to take a vertical shot and "tells" the camera to display it that way later when viewed on the TV screen during playback.

O S - stands for Operating System. It is the architectural foundation of every computer and therefore an absolute necessity for a computer to display information or software content.

OSI - open system interconnect

overexposure - An image that appears too light. All the highlights and colors are totally lost and usually unrecoverable even with sophisticated software.

P

PAL - Phase Alternation Line. The 50 field video format used primarily in Europe except France, former east block countries and northern regions of Africa.

palette - A thumbnail listing of all available colors to a computer or other devices. The palette allows the user to chose which colors are available for the computer to display. The more colors the larger the data and the more processing time required to display your images. If the system uses 24-bit color, then over 16.7 million colors are included in the palette.

Palm Pilot - a handheld device, capable of storing data and connecting to the Internet.

pan - aka panning. The horizontal movement of a camera, following an object.

panorama - Capturing a series of images to create a picture wider than what you could capture in a single image. Requires special software to combine and blend the images into one finished image.

parallax - An effect seen in close up video and photography where the viewfinder does not see the same as the lens due to the offset of the viewfinder and the lens. This is not applicable if using a LCD screen or monitor as a viewfinder.

password - a string of characters or numbers that a user enters in today's network or banking environment to gain access to protected files or to receive certain privileged information.

Pb - petabit. A petabit is equal to 1,000,000,000,000,000 bits

PB - petabyte is a unit of information equal to 1,024 terabytes or one quadrillion bytes.

P C - 1) In computer terms it means Personal Computer as in IBM-PC. 2) In camera terms it is meant to describe a type of flash sync connector, popular on most film cameras.

PC Card - Refers to a credit card-sized device which can be a flash memory card, a network card, a modem or even a hard drive and can be inserted into laptops.

PCI - peripheral component interconnect. A so called PCI card will fit in slots on laptop computers to connect to other peripherals. Most commonly used for rapid transfer of data from a CompactFlash or SmartMedia type memory card to the host PC.

PDA - Personal Digital Assistant. It is a small, handheld wireless device for transmitting pages, data messages, faxes and e-mails. It also acts as an electronic organizer, giving you access to schedules and contact lists. The term is often used interchangeably with PIM.

PDF - permanent data file or permanent data format. Usually, a file in this format is a legal document or contract and is meant to be printed unchanged and "as-is".

Personal Digital Assistant - aka PDA

PGA - pin grid array. A square integrated circuit with connecting pins surrounding the bottom edges on all four sides. PGA is the form factor frequently used for microprocessor chips.

PGP - 1) pretty good privacy. 2) A controversial freeware program created in June 1991 by Philip Zimmermann, PGP is designed to encrypt data for security.

phone book - The collection of telephone numbers you have stored into your phone's internal memory.

PhotoCD -   aka. picture CD. A disk that contains images of your photographs which were processed by a professional service and then scanned using an expensive drum scanner. You can chose between several different sized resolution images of each of your film pictures, from small to very large. PhotoCD is multi-session which means more than one roll of pictures may be placed on each disc.

photo viewer - aka SD / PC card slot allows you to easily display still photos that you've captured onto your digital camera's SD Memory Card on many Plasma or LCD displays. Photos can be viewed as a slide show, or in photo-album format. You can also easily rotate images using the remote control.

PICT - A graphics file format used primarily on Macintosh computers. PICT files can contain both object-oriented and bit-mapped graphics. There are two types: PICT I and PICT II. PICT II is the current standard and supports color up to 24-bit.

PIM - 1) print image matching is a new standard of embedded color and printing information for digital cameras. Many of the camera manufacturers have joined with Epson and now embed the PIM information in the header of the JPEG images created. 2) Personal Information Manager aka. PDA

PIN - Personal Identification Number

PIN Code - Personal Identification Code.

ping - An acoustical signal or data stream that is send by one device to another to assure connectivity.

pixel - The individual imaging element of a CCD or the individual output point of a display device. This is what is meant by the figures 640x480, 800x600, 1024x768, 1280x960 and so on. Higher numbers are always better.

pixelization - The stair-stepped appearance of a curved or angled line in digital imaging. The smaller the pixels, and the greater their number, the less apparent the "pixelization" of the image. Also known as "jaggies".

Plug-n-Play - An automated installation process used in Computers to connect peripherals to a CPU. When new devices are plugged into the computer the operating system recognizes the device and prompts the user to choose setup options and finish installation.

PMD - passive matrix display. A type of liquid crystal display, that uses one transistor for each row of pixels and one for each column of pixels.

PNC - Personal Network Connection

PNG - Portable Network Graphics. An image file format. It is a compressed file format similar to JPG.

polarizer - A photographic filter for eliminating glare and reflections.

point and shoot - A term used for a simple, easy to use camera with a minimum of user controls. Generally the user turns the camera on, aims it at the subject and presses the shutter button. The camera does everything automatically.

POP - 1) Post Office Protocol is a protocol designed to allow single user hosts to read email from a server. There are three versions: POP, POP2 and POP3. 2) Point Of Presence.

POTS - Plain old Telephone Service. A term used in videoconferencing in which video and audio signals are transmitted via modem through regular analog phone lines.

post scripting - Allows recalling a number and adding more digits from the keypad before placing a call to keep repetitive number dialing faster.

post script file - a special formatting procedure for word processing files, that "locks" a document to avoid changes like auto reflow and re- formatting.

postmaster - On the Internet this is the administrator responsible for resolving email problems, answering inquiries and other related duties at a site.

PPI - Pixels Per Inch. A measurement to describe the size of a printed image. The higher the number the more detailed the print will be.

PPP - Point-to-Point Protocol. It provides a method for transmitting packets over serial point-to-point links.

Pre-Flash - Some digital cameras use a low-power flash before the main flash to set the exposure and white balance. This does not allow the use of a normal photo slave strobe as it will be triggered by the pre-flash.

Programmed AE - the camera picks the best shutter speed and aperture automatically, also called "Automatic" or "point-n-shoot" mode.

progressive scan - Term used to describe an image sensor that gathers its data and processes each scan line one after another in sequence.

progressive cinema scan - Achieves exceptionally faithful movie reproduction through a sophisticated analog-to-digital conversion. Movies on film are converted to NTSC interlaced video (480i) for television by a process known as telecine conversion, in which the 24 frame-per-second film is converted to video at 60 fields per second. The fields are then paired to create 30-frame-per-second 480i video. Some of the frames will contain dissimilar pairs of fields derived from two different frames of the original film, which causes artifacts. The progressive cinema scan circuitry converts 480i video to 480p while restoring the original frames of the film for a more authentic movie reproduction.

protocol - A formal description of message formats and the rules two computers must follow to exchange those messages. Protocols can describe low-level details of machine-to-machine interfaces or high-level exchanges in which two programs transfer a file across the Internet.

private mode - will give the caller a one-to-one communication between two individuals through a portable device.

prosumer - Is a term used to describe more expensive semi-professional camcorders and cameras. The average prosumer unit has some professional features, but is made for the consumer market and costs about $1,000.

PSTN - public switched telephone network.

purple fringe effect - aka. chromatic aberration.

Q

query - To mark an item, an action or command with a notation in order to question its validity or accuracy. For instance you can query a cell in a worksheet, query a print command and query the printing on a printer itself.

quick access menu - this function allows you to tailor your phone by programming your most frequently used features into your personal menu. Access these features with just two quick key presses.

QuickTime - A motion video standard created by Apple. They have an entire QuickTime web site to explain it. QuickTime video sequences can contain an audio track and are stored as .MOV files.

Quiet Time - a function on cellular phones that allows you to select a time period during which the pager receives messages and performs all functions but without any audible or visual indication.

QVGA - Refers to a Quarter-VGA resolution (320 x 240) motion video sequences.

R

RAM - Random Access Memory . The most common type of computer memory. It is the place where the CPU stores software, programs, and data currently being used. RAM is usually volatile memory, meaning that when the computer is turned off, crashes, or loses power, the contents of the memory are lost. A large amount of RAM usually offers faster manipulation or faster background processing.

rangefinder - The viewfinder on most cameras is a separate viewing device that is independent of the lens. Often mounted above and to the right or left of the lens. It creates a problem known as parallax when trying to frame subjects closer than five feet from the camera, so it is advisable to use the color LCD when shooting close-ups for this very reason.

ratio - Is the relation in degree or number between two similar things.

RAW - In computer terms its transferring the raw unprocessed data - at 12 bits per channel - from the camera's imaging chip to your computer. Lossless compression is applied to reduce the file size without compromising any quality.

recall - This feature allows scrolling through the directory by name or by number according to user preference.

red-eye - An effect caused by an electronic flash reflecting off of the human eye and making it look red.

Compact cameras with the flash located close to the lens suffer the worst from this problem. This disturbing effect can be avoided by using a bracket to hold an external flash unit above and off to the side of the lens.

red-eye reduction mode - A special flash mode whereby a pre-flash or a series of low-powered flashes are emitted before the main flash goes off to expose the picture. This causes the pupil in the human eye to close and helps eliminate red-eye.

Region Code - An imbedded security code on pre-recorded DVD's, to avoid exchange from Country to Country in addition to the three different major TV standards PAL, NTSC & SECAM. Nine different Region Codes are currently used:

<u>*additional information:*</u>

REGION 1: USA, Canada

REGION 2: Japan, Europe, South Africa, Middle East, Greenland

REGION 3: South Korea, Taiwan, Hong Kong, Parts of South East Asia

REGION 4: Australia, New Zealand, Latin America including Mexico.

REGION 5: Eastern Europe, Russia, India, Africa

REGION 6: China

REGION 0 or REGION ALL: Discs are without any coding and can be played worldwide. However, PAL discs must be played in a PAL-compatible DVD players and NTSC discs must be played in an NTSC-compatible units.

reminder beeps - If the service provider offers Digital Message Access, the phone can be set to remind you that there are messages waiting to be viewed.

remote login - Operating on a remote computer, using a protocol over a computer network, as they are locally attached.

render - The final step of an image transformation or three-dimensional scene through which a new image is refreshed on the screen.

reset - An action that restores original manufacturer's settings, such as re-setting a clock, re-setting a computer or re-setting a printer.

resize - Usually means to take a large image and downsize it to a smaller one. Most graphic viewing and editing programs offer a resize option for this purpose.

resolution - The quality of any digital image, whether printed or displayed on a screen, depends in part on its resolution, made up of the number of pixels used to create the image. More and smaller pixels adds detail and sharpens edges.

RF - 1) Radio Frequency….a frequency in kilohertz or megahertz that carries an audio signal. 2) Range Finder - a type of camera viewfinder that uses one lens to frame your subject and another lens to capture the image.

RGB - Means red, green and blue - the basic colors from which all other colors are derived.

roaming - A service offered by most cellular service providers that allows subscribers to use cellular service while traveling outside their home service area. When they are outside their home service area and come within range of another cellular system, the ROAM indicator on the cellular phone will light to show that they are in range.

ROM - Read Only Memory. That is computer related storage space that cannot be altered.

ROTFL - Rolling On The Floor Laughing.

RS-232 - Standard type of serial data interconnection available on older PC type computers. It is also the slowest way to transfer image data from a camera.

S

saturation - The degree to which a color is undiluted by white light. If a color is 100 percent saturated, it contains no white light. If a color has no saturation, it is a shade of gray.

scanner - An optical device that converts images, such as photographs, into digital form so they can be stored and manipulated on computers. Different methods of illumination transmit light through red, green and blue filters and digitize the image into a stream of pixels.

scan rate - describes the fastest possible rate at which a server can update an item. The default value is 0, which indicates that the scan rate is not known. The scan rate may not be attainable by the server due to network load, server load and other factors.

scart - a multi-pin input/output connection for home video on European televisions and video equipment, carrying all possible signals between equipment.

scratch pad memory - An automatic memory feature that allows entry of a number into the keypad during a conversation for recall after the conversation has ended.

scroll keys - Can be found on a cellular phone and allow user to scroll forward and backward through menu options.

SCSI - A high-speed input/output bus used to "chain" computer hard drives, scanners, printers and other peripherals with each other. The chain requires to assign a different ID number to each piece of equipment.

SD - Secure Digital. Refers to the kind of memory card in image capturing equipment.

SECAM - Sequentiel Couleurs à Mémoire. A term used to describe the 50 field video output on televisions and video equipment. This standard is used in France and former French Colonies as well as a variety of Countries in the former Soviet Union.

sepia - The "brownish", mono-toned images from the "good ole days", now often found as a special image effect on some digital camcorders and cameras.

serial port - Same as RS-232.

server - A Company that provides users with Network access via Internet, cellular and satellite.

SFX - abbreviation for 1) special effects 2) sound effects

shell - The user interface to an operating environment.

SMS - Short Message Service. With this feature you can receive alphanumeric messages on your phone.

shutter - The gate that opens and closes in front of the lens to let light from the scene strike the image sensor.

shutter lag - The time between pressing the shutter and actually capturing the image. This is due to the camera having to calculate the exposure, set the white balance and focus the lens.

shutter priority AE - The user chooses a shutter speed and the aperture is automatically determined by lighting conditions. Shutter speed priority is used to control motion capture. A fast shutter speed stops fast action, a slow shutter speed blurs a fast moving subject.

signal strength - The signal strength meter is a visual indication of the relative strength of the cellular signal to help ensure that quality calls can be placed.

signature - The three or four line message at the bottom of an email message, identifying the sender.

silent keypad - The phone can be set to silence, so no tones can be heard when pressing the keys.

silent ringer - When this feature is activated the phone alerts an incoming call by flashing "CALL" in the display.

SIM - Subscriber Identification Module.

SIM Card - The Subscriber Identification Module card is a vital component in GSM operation. The user can store all relevant data for the phone on a removable plastic card. The card can be plugged into any GSM compatible phone and the phone is instantly personalized to the user.

SIM lock - is a series of programmed code numbers in a cellular device and usually only accessible to cellular phone companies and signal providers, to prevent a carrier switch by the user.

SIP - Is a telephone service that allows you to make phone calls worldwide using the Internet and a supplied IP.

SLIP - A protocol used to run IP over serial lines, such as telephone circuits or RS-232 cables, interconnecting two systems.

SLR - Single Lens Reflex. It means the camera has one lens that is used for both composing the frame and capturing the image to memory.

SmartMedia - a flash memory card. There are two types: 3.3v and 5v.

smoothing - Averaging pixels with their neighbors. It reduces contrast and simulates an out-of-focus image.

SMTP - Simple Mail Transfer Protocol. A protocol used to transfer electronic mail between computers. SMTP is a server to server protocol and other protocols are used to access the messages.

SNMP - Simple Network Management Protocol

SonicPath Audio System - Is new surround sound system technology, which provides thrilling audio to accompany high definition video. The four speakers in the SonicPath system deliver crystal clear high definition sound. The speakers are specially positioned on either side of the screen for maximum stereo separation. The positioning also provides a direct sound path to the viewer to maximize the home theater experience.

spam - A term referring to the act of posting the same message to several inappropriate newsgroups, or mass-mailing email messages to several users.

speakerphone - Enables the user to receive phone calls hands-free or conduct conference calls.

speed dialing - The touch of one or two keys on a phone, which are linked to a memory location inside the base or main board, provides rapid call initiation.

spike - An abnormally high, brief voltage fluctuation that can occur on an ordinary electrical line providing power to a facility. Spikes can damage electrical circuits and components.

spot metering - The camera's auto exposure system is focused on a very small area in the center of the viewfinder to critically adjust the overall exposure value ONLY for that area.

sprockets - is the term used to describe the tiny transportation holes on one or both edges of film material.

standby time - Is the time a battery can power a phone in the "waiting mode", ready to make or receive a call, without being used for an actual call. The longer a phone is in standby mode, the less standby time remains in the battery.

string - a series of data values, or bytes, which are usually character or numerical sequences.

surfing - A term for exploring the Internet, aka. "surfing the 'net". Most often used in reference to accessing sites on the World Wide Web.

Surround sound - While the term "surround sound" technically refers to specific multi-channel systems designed by Dolby Laboratories, it is more commonly used as a generic term for theater and home theater multi-channel sound systems.

S S F D C - Solid State Floppy Disc Card aka. SmartMedia card.

SVCD - Super Video Compact Disc is a CD-ROM disc that contains video and audio. Typically, a SVCD can hold about 35~45 minutes (650MB) of video and stereo-quality audio (depends on the data rate used for encoding). SVCD video is a slightly better quality than VHS video and the video and audio streams are stored in MPEG2 format, which have no TV system designation and no Region Code unlike a DVD.

additional information:

SVCD Video Parameter Settings
Frame Size: 480x480 (NTSC) or 480x576 (PAL)
Frame Rate: 29.97frames/second (NTSC) or 25 frames/second (PAL)
Video Data Rate: Variable bit rate up to 2600 kbps
Audio Settings: 32~384 kbps MPEG-1 Layer 2 audio bit rate

SVGA - Refers to an image resolution size of 1024 x 768 pixels.

SVHS - Super Video Home System. A video format delivering over 400 lines of crisp resolution by dividing signal streams into separate entities. In the 80's and 90's it was the format of choice for thousands of event videographers and small town Public Access formats.

sync - A special mode in digital still cameras that opens the shutter for a longer than normal period and fires the flash just before it closes. It is used for illuminating a foreground subject yet allowing a darker background to also be rendered. Good for night time shots of buildings with people in the foreground.

SysOp - The person responsible for maintenance of a given computer system. Short for "System Operator".

T

T1 - Two unshielded twisted-pair telephone lines are used to transport signals at 1.544 megabits per second, one pair for each direction. T1 equipment operates in full-duplex mode, separating the sent and the received signals at each end with components. The transmitted signal consists of pulses a few hundred nanoseconds wide, each inverted with respect to the preceding one. At the sending end the strength of the signal is 1 volt, and at the receiving end the strength must be greater than 0.01 volts.

T3 - A term for a digital carrier facility used to transmit a DS-3 formatted signal at 44.746 megabits per second.

T9™ - is a text input application that allows you to type messages with just one key press per letter on wireless phones. Much easier text input than traditional multi-tapping. Also called iTAP.

TA - Terminal Adapter. A device that adapts ISDN Basic Rate Interface channels to RS-232 and to other installed terminal equipment standards. TA connects to a computer or a router, replacing a modem.

TAC - Terminal Access Controller. A device that connects terminals to the Internet, typically with dialup modem connections.

TACACS - Terminal Access Concentrator Access Control Server is a protocol, which provides secure access to servers and data for LANs and telephone systems.

tag - A character or a string containing identifying information and connected to a piece of data, often at the beginning and the end. In HTML, a tag is placed before and after a text string to identify features or attributes of the data.

talk time - On cellular phones it is the total time a battery can power a phone for cellular calls. As the phone drains the battery during a call, the talk time left in the battery diminished until the phone is turned off or the battery is recharged.

T A P - Telocator Alphanumeric Protocol. is a half-duplex, ASCII-based protocol capable of submitting alphanumeric messages.

Tb - terabit. In measuring data transmission speed, a terabit is one trillion binary digits, or 1,000,000,000,000.

TB - terabyte. A unit of data equal to a trillion (1,099,511,627,776) bytes or 1024 gigabytes.

T C I T - Technical Committee of International Technology

T C P - Transmission Control Protocol. An Internet standard transport layer protocol. It is connection-oriented and stream-oriented, as opposed to UDP.

TCP/IP - Transmission Control Protocol over Internet Protocol. This is a common shorthand which refers to the transport and application protocols which run over IP. When your Internet connection fails, you may want to check the TCP/IP settings on your computer.

TDMA - Time Division Multiple Access is a digital format that divides a sequence of conversations into packets of data according to time.

TDS – 1) Time and Date Stamp. A marker that displays the time and date of an incoming message. 2) Thermal Dye Sublimation; aka Dye Sub. Meant is a high resolution continuous tone color printer. This technology allows the dot intensity to vary and to create many more colors than thermal wax. The dyes are vaporized at high heat and diffused across a small gap to the paper or transparency. Semi-transparent dots of cyan, magenta and yellow of varying intensities (usually 256 intensities) are overprinted to create more than 16 million hues. Thermal dye printers require special ribbon and paper.

telecine - a film transfer unit that enhances the quality of film material during its transfer by reflecting the image sequence through mirrors and prisms.

telephoto - The focal length that gives you the narrowest angle of coverage, good for bringing distant objects closer.

telnet - The standard Internet protocol for remote terminal connection service.

TFT - thin film transistor. This technology refers to the type of high resolution color LCD screen used in laptops, digital cameras or camcorders.

thread - A series of articles on the same topic.

throughput - meant is the output relative to input. In other words it's the amount passing through a system from input to output.

thumbnail - A small, low-resolution version of a larger image file that is used for quick identification.

TIA - 1) The Internet Adapter is a product that emulates a SLIP or PPP connection over a serial line. 2) used informally as an abbreviation for "thanks in advance".

TIFF - Tagged Image File Format. An uncompressed image file format that is completely lossless and produces no artifacts as it is common with other image formats like JPG.

tilt - a camera motion that captures or follows an object by moving from top to bottom or bottom to top.

time-lapse - Capturing a series of images at preset intervals.

timeline - 1) the chronological order in which things happened. 2) in non-linear editing the "assembly line" to place the movie clips.

TiVo - A digital recording device, allowing multiple recordings at the same time. A monthly subscription or one time payment for the life of the TiVo box is necessary to obtain service.

TLA - Three Letter Acronym

Transceiver - A radio transmitter and receiver combined into a single unit.

tri-band - Refers to cellular phones that operate on any of the three digital GSM frequencies-typically, 900 MHz, 1800 MHz and 1900 MHz-to increase service coverage. 1800 MHz is a GSM digital network used in Europe. This frequency allows for global roaming, where roaming agreements are in place.

tri-mode - Are cellular phones that offer dual-mode and dual-band capability to increase opportunities for obtaining cellular service. Dual-mode allows the phone to access digital and AMPS (analog) channels. Likewise, dual-band enables a phone to operate on two digital frequencies.

trojan horse - A computer program which carries within itself a means to allow the program's creator access to the system using it. 2) One of the first major viruses discovered on the Internet

troll - A term used to define a public message that is posted for the sole purpose of offending people and/or generating an enormous flood of non-topic replies.

true color - Color that has a depth of 24-bits per pixel and a total of 16.7 million colors.

trucking - In the film and video Industry it is a wheeled platform, sometimes equipped with a motor, that has a tri-pod and camera mounted on it. It has enough room to a camera operator and can smoothly follow horizontal action shots to create a life-like scene.

TrueSync® - Is software that allows your wireless phone or pager to exchange information with a compatible computer, PDA or other wireless device. Using a data (USB) cable, you can transfer contact information from your phone to your desktop, laptop or handheld computer, and vice versa.

TTFN - Ta Ta For Now

TTL - Through the lens, is a term used when talking about either an autofocus or auto exposure system that gather's its setting through the camera's lens.

TWAIN - Protocol for exchanging information between applications and devices such as scanners and digital cameras.

Type I, II, III – Refers to various PC ATA storage devices both flash memory and removable hard disk drives. Type I and II fit in the single-height card slots, Type III only fit in the double-height slots.

U

UDP - Underline{U}ser Underline{D}atagram Underline{P}rotocol. An Internet standard transport layer protocol. It is a more troublesome delivery service as opposed to TCP.

unanswered call indicator - When the phone is on it will alert and then keep track of calls not answered.

underexposure - A picture that appears too dark because insufficient light was delivered to the imaging system. Opposite of overexposure.

UNIX - a highly advanced multi-user operating system.

unsharp masking - A process by which the apparent detail of an image is increased; generally accomplished by the input scanner or through computer manipulation.

URL - Uniform Resource Locator. A means of identifying an exact location on the Internet.

USB - Universal Serial Bus. Meant is the hi-speed data port on digital cameras, camcorders and most PC and Mac computers. Faster than serial port or parallel port.

USB 2.0 - The newest USB standard, in which the speed almost matches the throughput speed to FireWire connection.

User-ID - A compression of "user identification". It always proceeds the @ sign in an email address.

username - A username consists of 1 to 8 characters, and only uses numbers 0 through 9 and the 26 alphabet letters. Usernames do not have spaces.

UUCP - UNIX-to-UNIX Command Protocol. a term that is commonly used to describe the large international network which uses the UUCP protocol to pass news and electronic mail.

UXGA - Refers to an image resolution size of 1600 x 1200 pixels.

V

VCD - Video Compact Disc. A CD-ROM disc that contains video and audio. Typically, a VCD can hold about 74 minutes (650MB) of video and stereo-quality audio. The video and audio are stored in MPEG-1 format and follow certain standards. VCD video quality is roughly the same as VHS video.

additional information:

VCD Video Parameters Settings

Frame Size: 352x240 (NTSC) or 352x288 (PAL)

Frame Rate: 29.97 frames/second (NTSC) or 25 frames/second (PAL)

Video Data Rate: 1152 kbps

Audio Settings: Stereo, 44.1kHz and 224kbps audio bit rate

V-chip - a technology that allows the user to block the display of television programming based upon its rating.

VCR - Video Cassette Recorder. A device that allows the recording of sound and picture signals to magnetic particle tape.

video - A videocassette or videotape, containing a recording of a movie, music performance, or television program.

Video-8 - A Sony© video format. Very popular because of its small size. The videotape only measured 8mm high.

VFD - Vacuum Fluorescent Display. A display that retains visibility in direct sunlight and is highly visible in darkness. It can be seen without distortion over a wide range of viewing angles and remains fully operational over a broad temperature range.

VGA - Refers to an image resolution size of 640 x 480 pixels.

VITC - Vertical Interval Time Code. It is time code information in digital form, added into the vertical space, aka blanking, of a video signal. This can be read by the video heads from tape at any time pictures are displayed, even during jogging and freeze but not during spooling.

VHS - Video Home System. Developed by JVC in the late 70's and a strong opponent of the simultaneously introduced Betamax format by Sony. The two systems were the first audio and video recording formats for home use, recording on a half inch magnetic tape with a running time of two hours and more.

VHS-C - Video Home System Compact. A smaller version of the VHS format, developed to downsize the bulky camcorder technology. A separate adapter is necessary to play VHS-C type cassettes in regular VCR's.

VibraCall™ - An alert feature on cellular phones in situations where ringing may not be appropriate, or in noisy places where ringing may not be heard, the phone can be set for discreet vibration.

video out - A connection with the ability to output its images to television screens and monitors using either NTSC or PAL format.

viewfinder - The eye level device you look through on camcorders or cameras to compose the image.

virus - A program which replicates itself on computer systems by incorporating itself into other programs which are shared among computer users.

voice mail - A feature that allows you to record or play back voice messages.

voice mail indicator - Your phone will indicate when you have received messages in your voice mail box.

volume control - allows you to adjusts volume levels for the earpiece, ringer, and speaker to your personal liking.

VRML - virtual reality modeling language. The standard for delivering virtual worlds on the Internet.

VSP - Vehicle speaker phone. It allows hands-free operation and a conversation to take place without using the handset. Enhances safety and convenience.

VTR - Video Tape Recorder.

W

W3C - <u>W</u>orld <u>W</u>ide <u>W</u>eb <u>C</u>onsortium. Meant is the body responsible for the technical evolution of the web. The W3C is now developing a technology to allow web access from non-PC devices.

wafer - A thin disk or a slice of silicon on which separate chips can be fabricated and cut into individual die.

WAIS - <u>W</u>ide <u>A</u>rea <u>I</u>nformation <u>S</u>ervers. A distributed information service which offers simple natural language input, indexed searching for fast retrieval and a "relevance feedback" mechanism, which allows the results of initial searches to influence future searches.

W A N - <u>W</u>ide <u>A</u>rea <u>N</u>etwork. The integration of geographically distant or technologically incompatible local area networks (LANs).

WAP - <u>W</u>ireless <u>A</u>pplication <u>P</u>rotocol. By working on top of standard data link protocols, WAP provides a complete set of specifications to support Internet protocols. Simply put, it's a special way of formatting content so that it can appear on small screens, like those on wireless phones or PDAs. WAP is becoming a global standard for developing and delivering Internet content to wireless devices.

W A T S - <u>W</u>ide <u>A</u>rea <u>T</u>elephone <u>S</u>ervice. A long-distance toll service, inward or outward, offered at a discount by a telephone company.

WAV - A Microsoft DOS and Windows file name extension used to indicate a sound file.

waveform monitor - A test instrument that displays a video signal graphed over a selected interval of time, rather than displaying a video picture.

WDM - <u>W</u>avelength <u>D</u>ivision <u>M</u>ultiplexing is the combination of more than one light source and a detector that operate at different wavelengths through the same fiber and that simultaneously transmit independent optical signals.

wet - Refers to an audio track that has been treated with signal processing, particularly reverb. A wet track is also a processed track with delay or other effects, whereas a "dry" track has no effects.

wet gate - A chemical process that coats the celluloid material, hopefully filling in digs and scratches or imperfections that occur in the negative to help restore the overall image.

white balance - Refers to adjusting the relative brightness of the red, green and blue components so that the brightest object in the image appears white.

white book - a government report, bound in white.

whois - refers to an Internet program which allows users to query a database of people and other Internet entities, such as domains, networks and hosts.

wide angle - The focal length that gives you the widest angle of coverage.

wideband video amplifier - a technology that expands the video signal frequency response to more than 10 MHz, allowing for an incredible horizontal resolution capability.

wide SCSI - Wide Small Computer System Interface. It is a form of SCSI-2 interface that allows a 16-bit bus that supports transfer rates up to 20 megabytes per second, as does fast SCSI.

Wi-Fi - 1) The Wi-Fi Alliance is a nonprofit international association formed in 1999 to certify interoperability of wireless Local Area Network products based on IEEE 802.11 specification. 2) freely translates into wireless fidelity. It actually refers to a set of wireless networking technologies more specifically known to us as 802.11a 802.11b and 802.11g. These standards are universally in use around the globe, and allow users that have a Wi-Fi capable device, like a laptop or PDA, to connect anywhere where there is a Wi-Fi access point available.

WiFi-SM - is an Internet connected wireless device that you can fix on any part of your body. It automatically detects the information from approximately 4,500 news sources worldwide updated continuously and analyses them looking for specific keywords.

WIMP - stands for <u>W</u>indows, <u>I</u>cons, <u>M</u>enus, and <u>P</u>ointing device; a graphical user interface originally used by the Mac OS, and replicated by other systems.

winsock - is a TCP/IP stack that allows you to use your modem to send data to/from the Internet.

WML - <u>W</u>ireless <u>M</u>arkup <u>L</u>anguage. A language that allows the text from web pages to be displayed on wireless phones and handheld devices. WML is part of the Wireless Application Protocol. With a device that has a WML-compliant mini-browser , you are able to browse Internet sites written in HDML.

worm - A computer program which replicates itself and is self-propagating. Worms, as opposed to viruses, are meant to spawn in network environments. WWW - Stands for world wide web, also known as Information Super Highway. Connects people and Companies around the world by using the Internet. It is a hypertext-based, distributed information system created by researchers at CERN in Switzerland. Users may create, edit or browse hypertext documents.

WYSIWYG - abbreviation for "<u>W</u>hat <u>Y</u>ou <u>S</u>ee <u>I</u>s <u>W</u>hat <u>Y</u>ou <u>G</u>et".

X

xD Picture Card - A new flash memory card standard that was co-developed by Fujifilm and Olympus in mid-2002. Rumored to be replacing SmartMedia which has stalled at 128MB. xD is scheduled to go as large as 8GB in a form factor the size of a postage stamp.

XGA - Refers to an image resolution size of 1024 x 768 pixels.

XLR - An audio connector used for most low-impedance microphones and for balanced lines between professional signal-processing equipment.

XML - Extensible Markup Language. XML documents are comprised of storage units called entities, which contain either parsed or unparsed data.

Y

Y/C - A type of video signal used in the Hi-8 and the S-VHS videotape formats. It transmits luminance and color as separate entities, avoiding the combination used in composite video and the resulting loss of picture quality.

YMCK - A common abbreviation for yellow, magenta, cyan and black. It usually refers to the color print cartridge in desktop printers.

Z

ZIP disk - A data storage disk in its own shell. Pre-runner of the CD-R.

ZIP file - A compressed archive file.

ZLR - <u>Z</u>oom <u>L</u>ens <u>R</u>eflex. A term created by Olympus to describe their fixed mount lens SLR type cameras. A SLR camera has an interchangeable lens, a ZLR has a non-removable lens.

zoom - To enlarge or to reduce the size of an image as it is displayed in a camera. This is done by changing the focal length of a zoom lens.

zoom lens - A variable focal length lens. The most common on digital cameras has a 3:1 ratio, i.e. 35-105mm.